CRYSTALLOGRAPHY

CRYSTALLOGRAPHY

R. Steadman
University of Bradford

VNR Van Nostrand Reinhold (UK) Co. Ltd.

© Van Nostrand Reinhold (UK) Co. Ltd., 1982

Reprinted 1983

Published by Van Nostrand Reinhold (UK) Co. Ltd.
Molly Millars Lane, Wokingham, Berkshire, England

Library of Congress Cataloging in Publication Data

Steadman, R.
   Crystallography

      1. Crystallography.   I. Title.
QD905.2.S73   548   81-24069
ISBN 0-442-30498-6 (pbk.)   AACR2

Printed in Great Britain by
The Thetford Press Limited, Thetford, Norfolk.

PREFACE

All students of materials science, physics, chemistry and metallurgy
meet the subject of crystallography, and it makes its appearance,
often as a brief, early chapter, in many textbooks on these subjects.
But a passing encounter with crystallography seems to be of no benefit
to a student, leaving him with little more than an aversion to
Miller indices and a feeling that his mind is not equipped for these
three-dimensional affairs.   There can be few subjects so simple
which have created so much bewilderment.

This book attempts to convince students that crystallography is based
on a few simple ideas which are very easily understood, and it does
it by presenting them briefly and following them with a succession of
short questions.   Most of the pages carry problems, and it is largely
a book of illustrations and questions with no lengthy exposition.

It is intended to be a book which every student can master from
beginning to end within the first term or so, if necessary, of a
university course, and precise aims have been kept in view.   These are
to provide a workmanlike knowledge of crystal geometry and the ability
to interpret x-ray powder photographs and electron diffraction patterns.
Each subject is approached in as direct a manner as possible in the
hope that the scientific ideas may soon appear to be no more than
common sense, and no suggestion is made that there may exist wider
issues beyond the student's present grasp.

It is not necessary to study the book systematically from the beginning,
and Section 2 makes a good starting point for students in a hurry.
Anyone taking an x-ray photograph within a week or two of arriving at
university will begin with Section 5 and leave the rest until later.

C O N T E N T S

SECTION   I

Structures and lattices

The variety of crystal structures is enormous, and we often need to
be able to talk about crystals in general and to discuss rules which
apply widely to all crystals rather than to one particular structure.
We are obliged to simplify the crystal, to strip it down to essentials
and reveal its geometry.   The atoms are swept away and leave behind
only a skeleton, a lattice as it is called, of mathematical points,
and each point replaces anything from one to several hundred of the
original atoms.

This first section introduces lattices and their relation to crystal
structures.

For convenience, two-dimensional lattices are discussed first, and we
find that, surprisingly, there are only five of them.   They are known
as square, rectangular, centred rectangular, oblique and hexagonal.
From these we move on to 3-dimensional lattices, of which there are
fourteen.

You will realise, before the end of this section, that the ruthless
stripping away of the atoms of a structure to leave behind the dead
lattice may make for simplicity, but it removes all trace of the real
crystal and almost all knowledge of its physical properties.   To know
the lattice of a crystal structure is to know very little.   Iron and
manganese, for example, both have the body-centred cubic lattice, but in
saying this we are giving no impression at all of the atomic arrangements,
since where iron has one atom manganese has twenty-nine.

Once we have completed the study of crystal geometry we shall, in
Section 4, replace the atoms and rebuild the crystal structure.

Imagine that this pattern is a crystal structure, a two-
dimensional structure, of large and small atoms.
We shall be using 2-D patterns to represent crystals in the
next few pages because they are easier to sketch than real
3-D crystals.

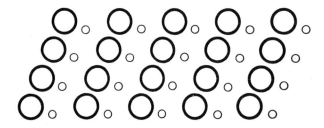

The LATTICE of this structure, as it is called, is the oblique
pattern of points given below.

Each point on the LATTICE represents a pair of atoms in the
STRUCTURE.

On the left is a structure (of atoms) with its square
lattice (of points) shown on the right.

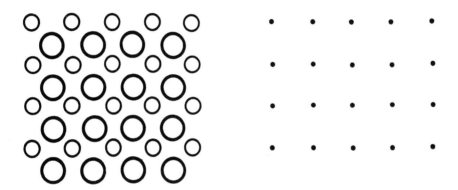

The next structure has a rectangular lattice, in which each
point represents three atoms of the structure.

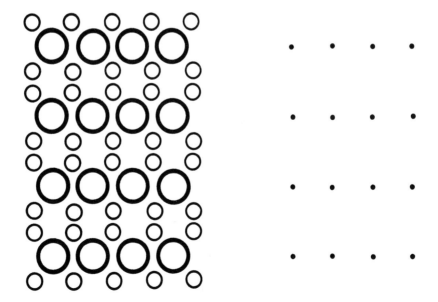

There are five structures on this page, each with one unit
of its lattice. The five lattices are shown in full on the
next page.

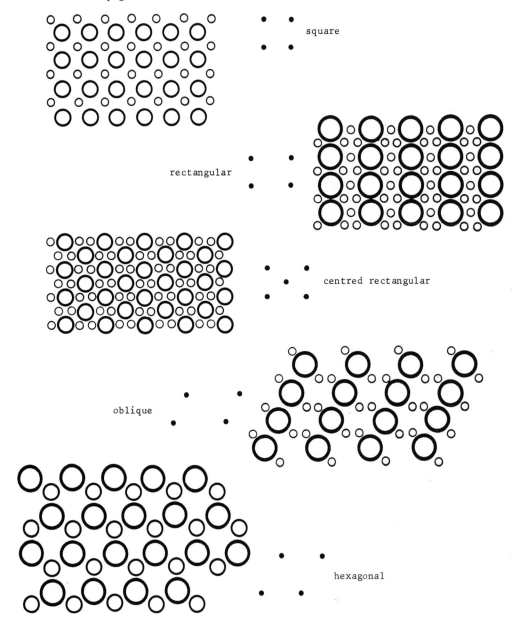

square

rectangular

centred rectangular

oblique

hexagonal

The five lattices mentioned on the previous page are the only ones which exist, and every 2-D pattern, whatever it is, has one of those five plane lattices. Here they are drawn out more extensively. Why there are five and no more will become clear later.

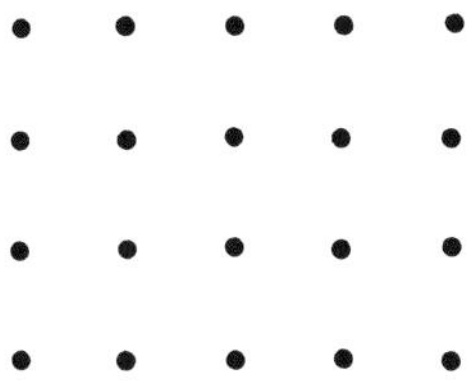

the square lattice

the rectangular lattice

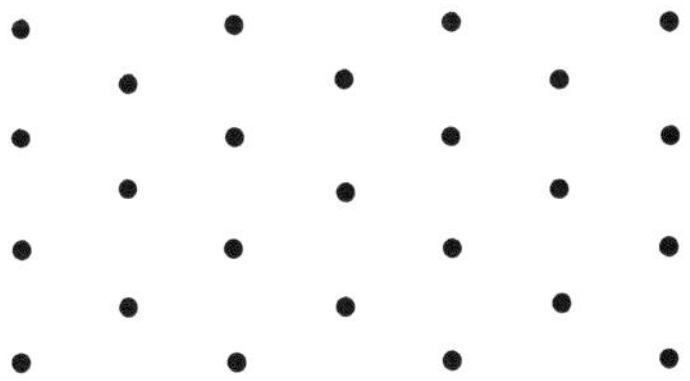

the centred rectangular lattice

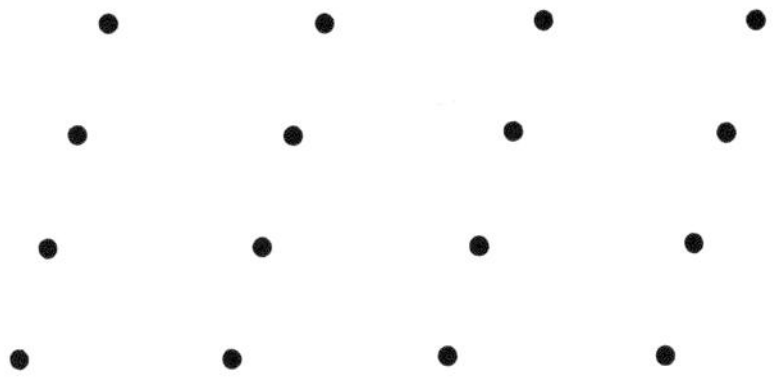

the oblique lattice

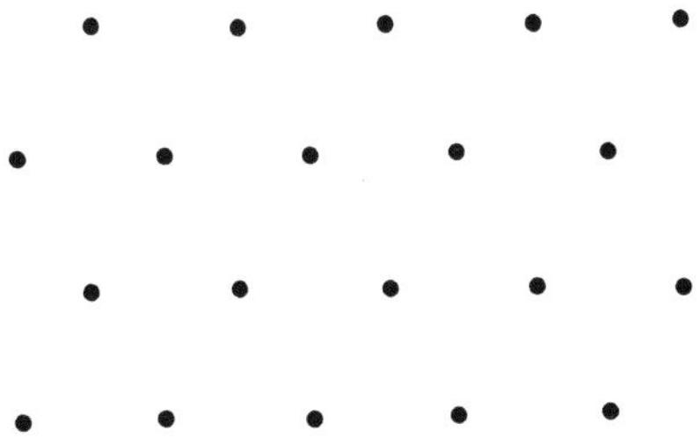

the hexagonal lattice

Try to identify each of the lattices shown below and on the next page, remembering, of course, that there are only five possibilities – square, rectangular, centred rectangular, oblique and hexagonal.

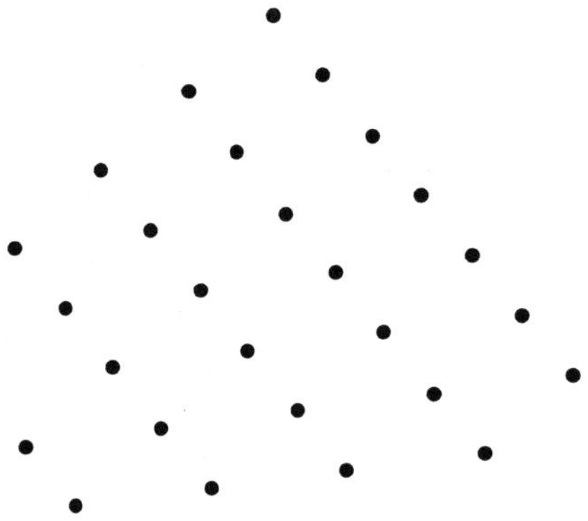

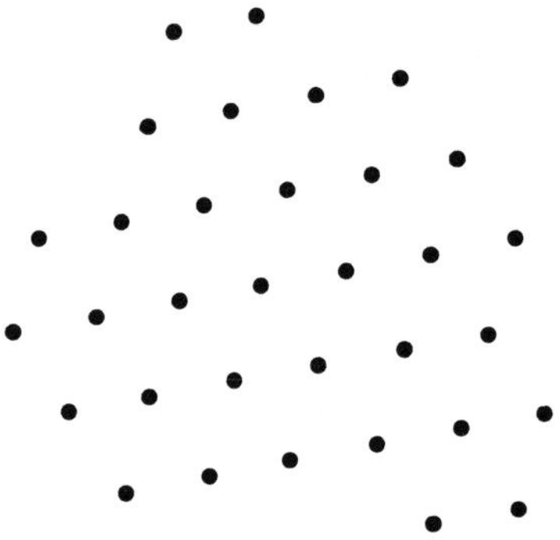

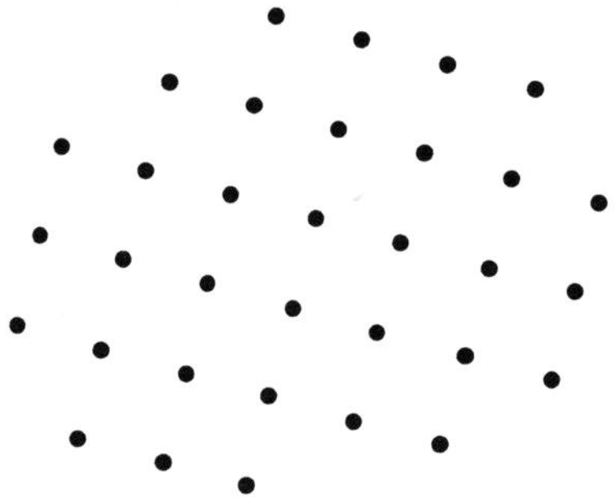

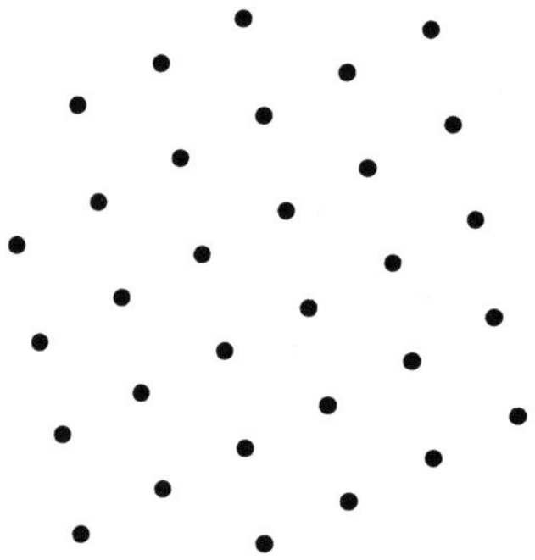

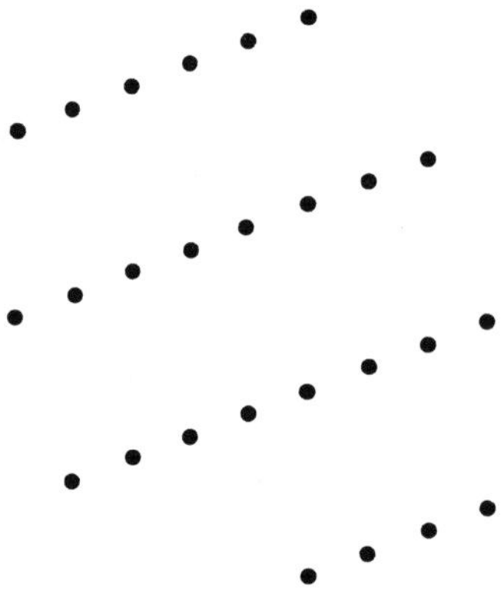

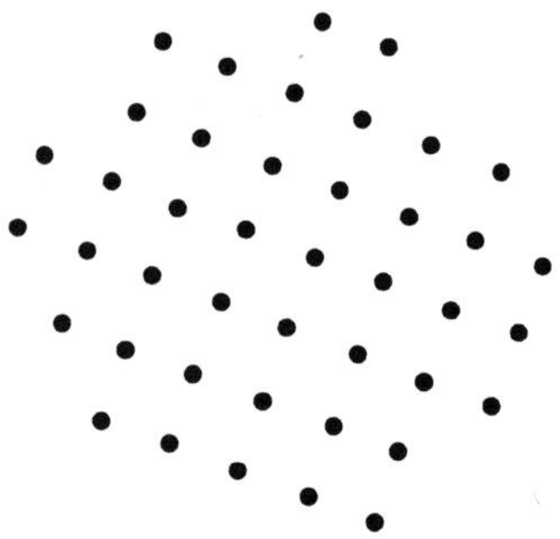

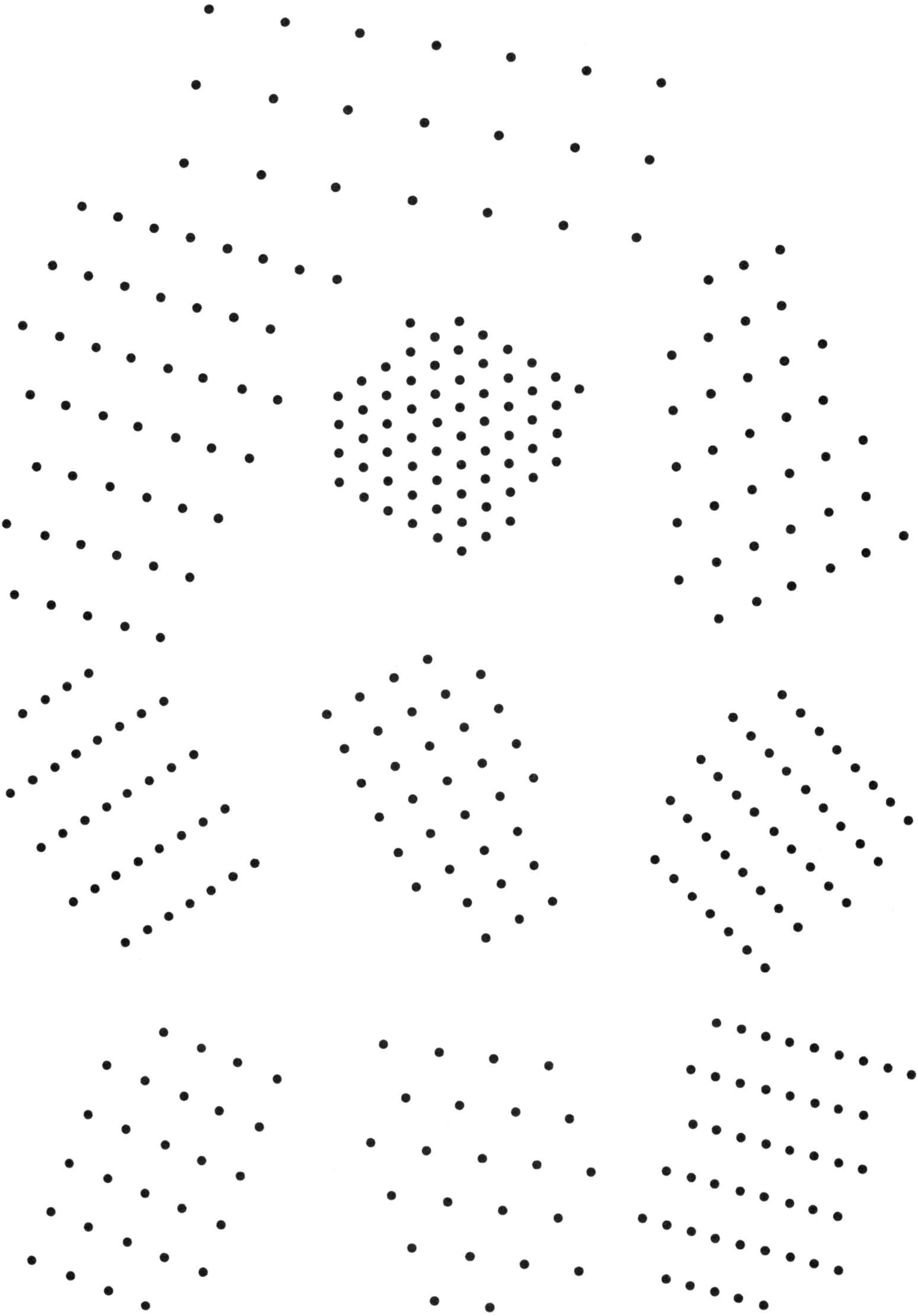

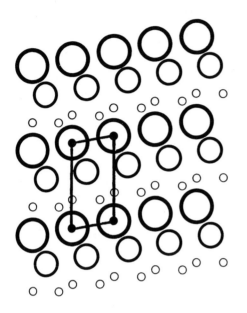

The lattice of this structure is oblique. Notice that each lattice point represents four atoms of the structure, because there are two atoms within the outlined cell and parts of six others, making a total of four. Since there is one lattice point per cell, there are four atoms per lattice point.

The structure on the left below has a centred rectangular lattice, and there are five atoms of the structure to each point of the lattice. Four atoms are within the cell and many others partially in it, making up a total of ten. Each cell, being centred, has two lattice points, so five atoms per lattice point. On the right is a structure with a rectangular lattice, and this time there are six atoms to each lattice point.

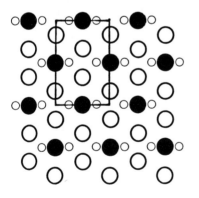
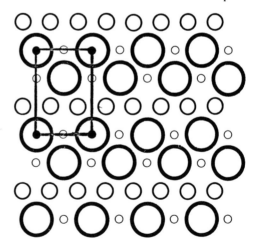

Identify the lattice of each of these crystal structures. (One of each of the five lattices is represented.) Find also the number of atoms which each lattice point represents.

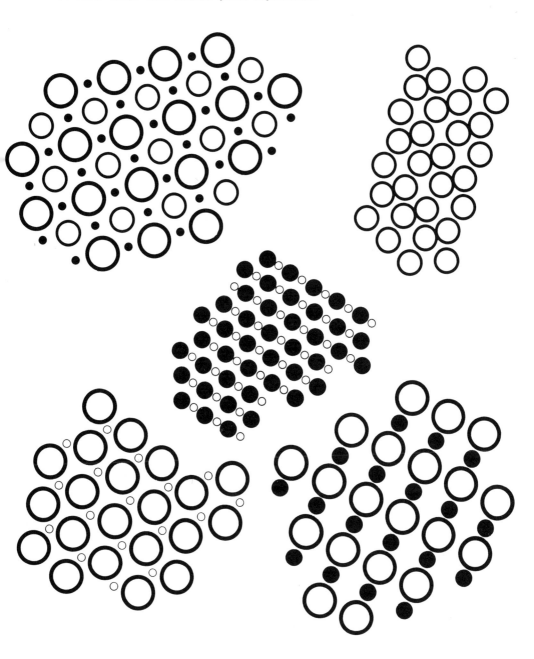

Again, as on the previous page, identify the lattice of each of these structures, and give the number of atoms which each lattice point represents.

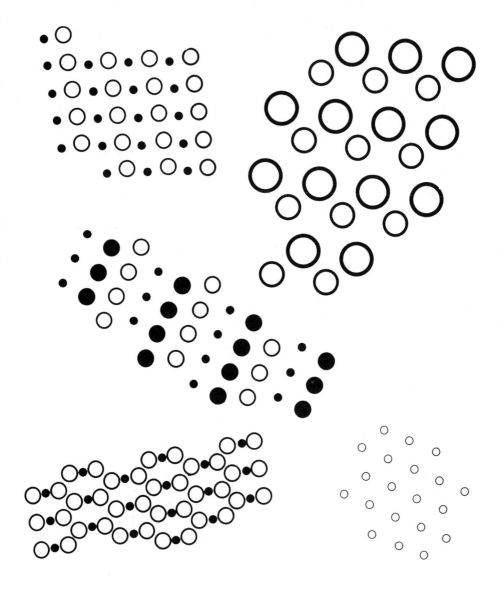

Notice how we never get away from the same five lattices, whatever structures we draw. No other lattices exist.

To understand this, try to devise a new lattice, a centred square lattice, say. We have met the centred rectangle, so why not a centred square?
We try it in the next diagram.

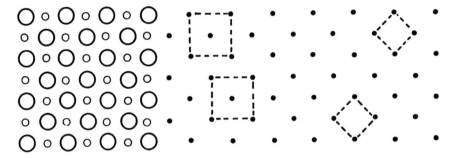

But what appears on the left to be a centred square lattice is clearly only a smaller square at 45°.

In the same way, try a centred hexagonal lattice. On the left below it is uncentred, and on the right it is centred, but you can see that centring has simply recreated the rectangular lattice.

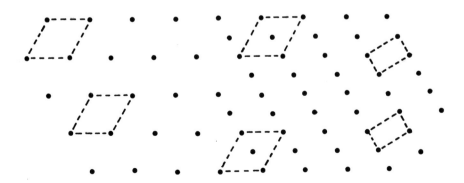

Couldn't we simplify matters by calling all lattices oblique,
even the square ones and the rectangular?   This is a reasonable question
to ask, and the diagrams below show the possibilities.

Yes, there is nothing to prevent our doing this, and we may call
all lattices oblique if we wish.

But the human brain has an instinct for symmetry and thinks of a
square as a square rather than just an unusual rectangle.
A rectangle too is seen as more than an exceptional parallelogram.

Every crystal structure has a certain symmetry, and each unit cell
(square, rectangular, oblique, hexagonal) has its own symmetry, so
we use the unit cell which has the symmetry which matches the crystal
structure.   By using a word such as "rectangular" to describe a lattice,
we are stating something useful about the symmetry of the crystal structure
which would be missing if we called it oblique.

We have so far used the term "lattice" without troubling ourselves
with a definition of it, but a definition is necessary, and
we give it here in two parts.

1.   A LATTICE IS AN INFINITE PATTERN OF POINTS.

2.   ALL POINTS HAVE THE SAME SURROUNDINGS IN THE SAME ORIENTATION.

Would you call this a lattice?   It is certainly a pattern of points,
but nothing like it appears among the five lattices we have
encountered.

Some of the points in it have these surroundings

and some have these

Clearly then, there are two kinds of point in this pattern.
So does it satisfy part 2 of the definition, and can it be a
lattice as we understand it?

We shall now look at the three-dimensional lattices, but
before we do this we need to recognise seven systems of
axes which give us the seven unit cells.

These seven systems of axes are called, rather oddly,
the seven "crystal systems".

The axes are naturally the x-, y- and z-axes, and along them
point the unit cell vectors $\underline{a}$, $\underline{b}$ and $\underline{c}$ which form the edges
of the unit cells.
The magnitudes of these vectors are a, b and c, the lattice
parameters.
On all the diagrams appear the letters a, b and c, and we
shall refer to the a-axis, b-axis and c-axis when strictly
perhaps x-axis, y-axis and z-axis would be correct, but the
meaning is quite clear and it makes the text match the
diagrams.

The first five systems, decreasing in symmetry.   Start with
the cube at the top, and distort it progressively as you go down
the page.

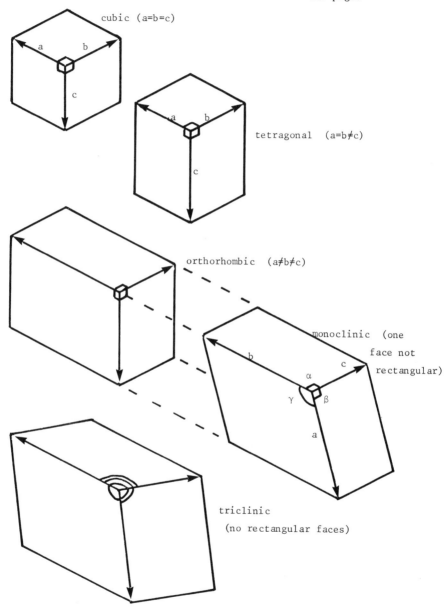

cubic (a=b=c)

tetragonal (a=b≠c)

orthorhombic (a≠b≠c)

monoclinic (one
face not
rectangular)

triclinic
(no rectangular faces)

These two don't quite fit into the sequence of decreasing symmetry formed by the five on the previous page.

The rhombohedral unit cell on the left is like a cube stretched (or compressed) along a diagonal.   It is also called the trigonal unit cell.

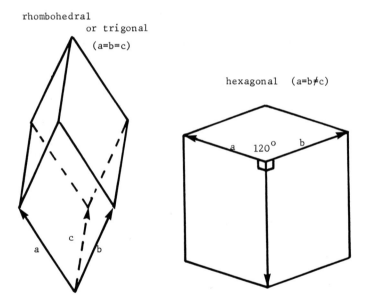

rhombohedral
   or trigonal
   (a=b=c)

hexagonal   (a=b≠c)

Notice that the hexagonal unit cell does not have a hexagonal face, but a group of them together give a hexagon-like pattern.

To become familiar with the seven crystal systems, complete the table below.

It gives data about the unit cells of some chemical compounds, but the table is incomplete, and you are expected to provide the missing data after studying the data already given.

a, b and c are given in angstrom units (Å) $(1\text{Å} = 10\text{m}^{-10})$

chemical formula	a	b	c	α	β	γ	system
Ti_5O_9	5.57	7.12	8.86	97.5°	112.3°	108.5°	
Al	4.05	4.05	4.05	90°	90°	90°	
Zn	2.66		4.95				hexagonal
V_2O_5	11.5	3.56	4.37				orthorhombic
NaSb	6.80	6.34	12.48		117.7°		monoclinic
Cu_2O			4.27				cubic
TiO_2	4.59		2.95				tetragonal
$CaCO_3$	6.36			46.1°			trigonal
Fe_3C	4.52	5.09	6.74	90°	90°	90°	
P_4S_{10}	9.07	9.18	9.19	92.4°	101.2°	110.5°	
Pb_2O_3	7.05	5.62	3.86		80.1°	90°	monoclinic
β-U	10.59	10.59	5.63	90°	90°	90°	
WO_3	7.27	7.50	3.82	90°	89.9°	90°	
MgO		4.44					cubic
Al_2O_3	5.13	5.13	5.13	55.3°	55.3°	55.3°	
$FeCl_3$	6.76			53.2°			trigonal
$AgNO_2$	3.53	5.17	6.17	90°	90°	90°	
H_2O	4.52	4.52	7.37	90°	90°	120°	
$CaCl_2$	6.24	6.43	4.20				orthorhombic
CaC_2		3.87	6.37				tetragonal

The 14 three-dimensional lattices   (The Bravais lattices)

Here now are nine of the only 14 ways of arranging points in space, sticking to our rules on page 13.

For each of them you are expected to imagine the pattern of points (not atoms, don't forget) repeated infinitely throughout space to make the complete 3-D lattice.

On this page, all faces are right-angled.

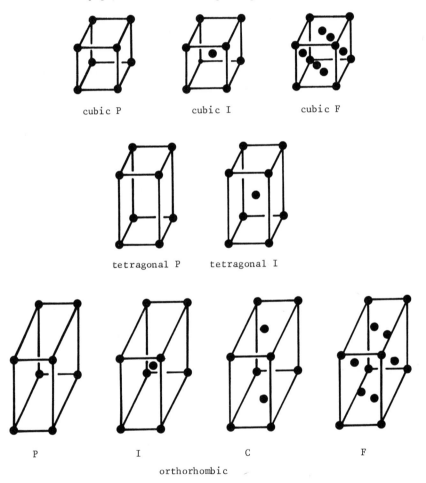

cubic P   cubic I   cubic F

tetragonal P   tetragonal I

P   I   C   F

orthorhombic

P means primitive.   I means body-centred.   F means face-centred.
C means centred on the C-face, the one perpendicular to the c-axis.

These are the remaining five of the fourteen lattices.

The symbol P for primitive indicates simply that the lattice is
not centred in any way, either at the body-centre or on its faces.
An exceptional symbol is R for rhombohedral (or trigonal), since
this lattice too is primitive.

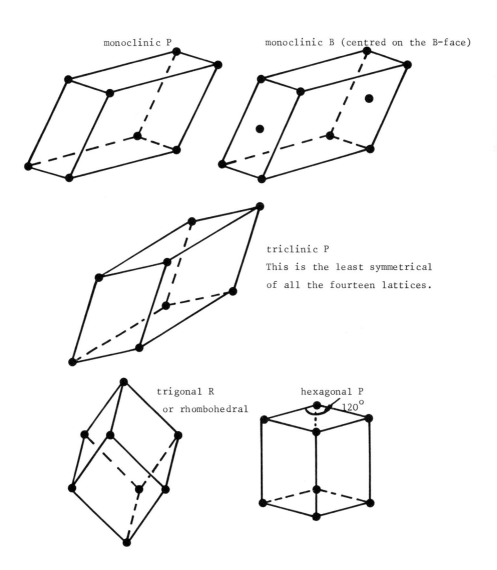

monoclinic P

monoclinic B (centred on the B-face)

triclinic P
This is the least symmetrical
of all the fourteen lattices.

trigonal R
or rhombohedral

hexagonal P
120°

Why aren't there more than fourteen 3-D lattices?
Answer this by attempting to devise some new ones, just as we did
when discussing the five 2-D lattices.

Attempt No.1

Take the tetragonal lattice and
face-centre it, just as we do
with the cubic.
Four unit cells of the lattice
are shown in the diagram, but you
can see that when looked at in a
different way it is the body-centred
tetragonal lattice which we have
already met.

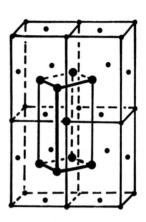

Attempt No.2

Centre the cubic lattice on one face only, as we do with the
orthorhombic (C) and monoclinic (B) lattices.
The diagram shows two cubes, each centred on the top (and bottom)
face.   But we now have, at $45°$, a tetragonal lattice, so again
nothing is new.

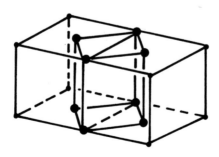

Attempt No.3

Try creating a new lattice by centring the triclinic lattice on its faces, or perhaps make it body-centred.

The attempt is obviously not worth making. Whatever lattice we produce, it will be possible to choose a unit cell for it in such a way that it is still a primitive triclinic lattice.

Couldn't we choose a triclinic unit cell for any of the 14 lattices, so that any lattice can be regarded as triclinic?

Of course we could, just as all the 2-D lattices could be looked at as oblique lattices, as we showed on page 12.

But in doing this we would be using a name which hides the lattice's symmetry. If a lattice is cubic it is highly symmetrical and it is necessary to know of this symmetry, and to say that it has a triclinic unit cell would be misleading. Sodium chloride, for example, has a well-known cubic structure with the cubic F lattice, and a unit cell could be chosen to describe it as a triclinic P lattice, but it would be an obtuse-minded crystallographer who did so.

A reminder to end this section

A crystal STRUCTURE

is made of ATOMS

A crystal LATTICE

is made of POINTS

A crystal SYSTEM

is a set of AXES

SECTION 2

Planes and directions in a crystal

For two purposes, we frequently have to think of planes of lattice
points extending through a crystal, not individual planes but sets
of planes all parallel to one another.

The first is that planes of atoms reflect the various radiations,
principally electrons, x-rays or neutrons, which are used to
investigate the crystal. We are interested both in the planes and
in the direction of the beam of radiation too, so it will be
necessary also to specify a direction in the crystal.

The other purpose is to describe how crystals become deformed
plastically (and we are thinking mainly, but not solely, of metal
crystals here). Planes of atoms slide over their neighbours, and
we need to be able to specify the slip planes, as they are called,
and the direction in which slip is taking place.

This section deals with identifying the planes and directions in a
crystal. Planes are known by their Miller indices, the integers
h, k and ℓ, and directions by their indices u, v and w, and it is
essential, given a set of planes for example, to be able to specify
the indices with ease and certainty.

How to determine the Miller indices (hkℓ) of a set of parallel
planes in a lattice.

1.   MOVE ALONG THE WHOLE LENGTH OF THE UNIT CELL VECTOR a.
     COUNT THE NUMBER OF SPACES CROSSED, AND THIS IS h.
     BY SPACES WE MEAN SPACES BETWEEN PLANES.

2.   DO THE SAME ALONG b TO GET k.

3.   DO THE SAME ALONG c TO GET ℓ.

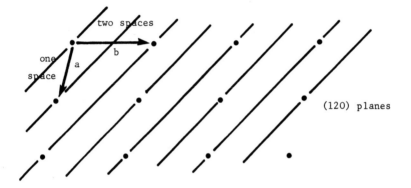

(120) planes

The c-axis is ⊥ to the paper.   You cross no spaces in going along it
because the planes are also ⊥ to the paper, so ℓ is zero.

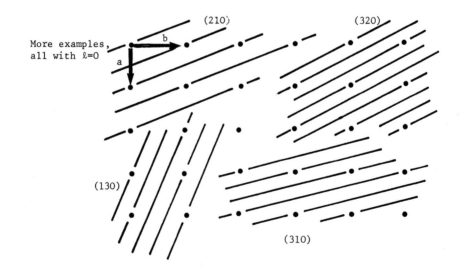

More examples,
all with ℓ=0

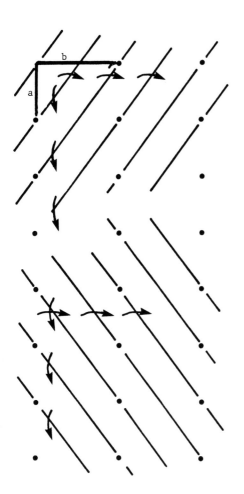

These planes are (120) planes.

The curved arrows are intended to show that in going along a or along b you cross each plane from the upper side to the lower side.

These planes are called either (1$\bar{2}$0) or ($\bar{1}$20) because in going along a you cross each plane from the upper to the under side, but going along b you cross from the under to the upper side.

The indices (1$\bar{2}$0) and ($\bar{1}$20) mean the same thing; there is absolutely no difference between them and you can use whichever you personally happen to prefer.

In the same way, the (120) planes at the top of the page could have been called ($\bar{1}\bar{2}$0). Always, the (hkℓ) planes and the ($\bar{h}\bar{k}\bar{ℓ}$) planes are the same set of planes. The examples given in the next few pages will help to make this clear.

Give the indices of all the unindexed planes below. Notice that the same monoclinic lattice covers the entire page, and the vectors <u>a</u> and <u>b</u> shown at the top apply to the whole page.

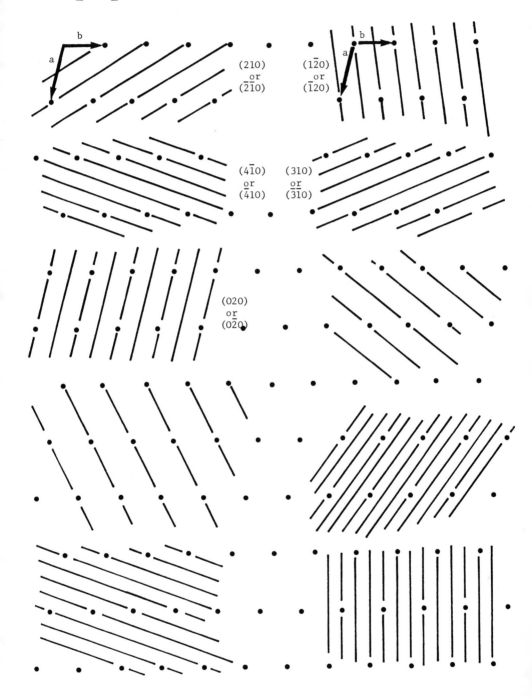

(210)
or
($\bar{2}$10)

(1$\bar{2}$0)
or
($\bar{1}$20)

(4$\bar{1}$0)
or
($\bar{4}$10)

(310)
or
($\bar{3}\bar{1}$0)

(020)
or
(0$\bar{2}$0)

Index all the sets of planes below.

This time an orthorhombic lattice covers the whole page, but of course the same method of determining (hkℓ) applies, whatever the lattice.

Be sure of vectors <u>a</u> and <u>b</u> before attempting to index the planes. If we were to use different axes the indices (hkℓ) would change too.

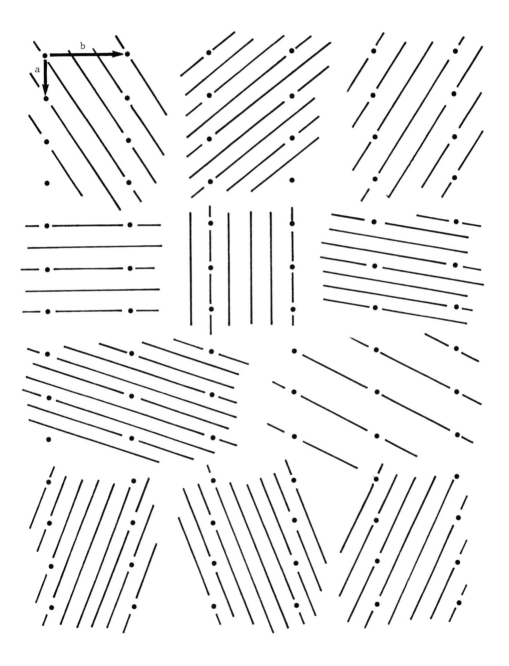

To show planes in three dimensions we have used only one unit cell in each drawing to avoid confusion.

As before, to find h, k and ℓ, move along <u>a</u>, then <u>b</u>, then <u>c</u> (in the positive direction of course), and count the spaces.

We have used the same axes in all the drawings.

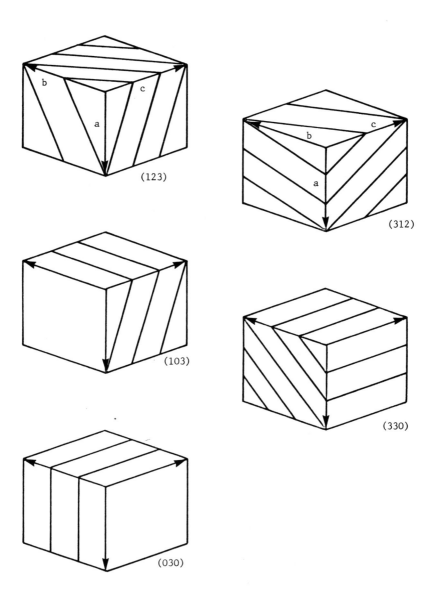

h, k and ℓ for the planes in the top drawing are all positive (or, if you wish, all negative).

The axes pierce the planes like needles piercing pieces of cloth, and all three needles go through the cloth from the same side, which is the upper side as we see it here.

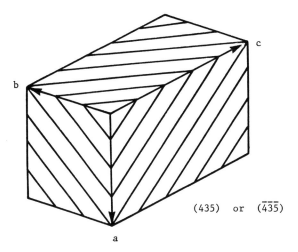

(435)   or   ($\bar{4}\bar{3}\bar{5}$)

But in the next drawing the a-axis needle pierces the cloth from one side and the b-axis and c-axis needles pierce it from the other.

So we make h positive and k and ℓ negative, or the other way round if you wish.

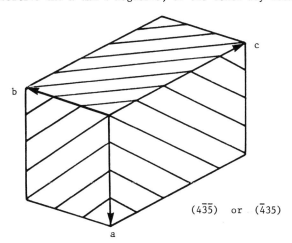

(4$\bar{3}\bar{5}$)   or   ($\bar{4}$35)

Index all the planes shown.

The indices of the planes on the left are all positive (or all negative) and those on the right are mixed.

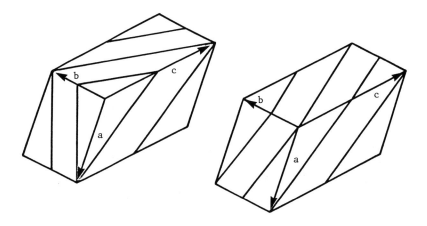

The planes below are also two related sets. Remember that when the axis runs parallel to the planes the index is zero.

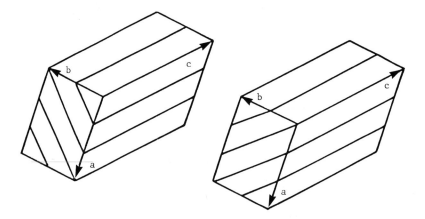

Now index all these planes.

For the whole page, use the axes
given in the top two unit cells.

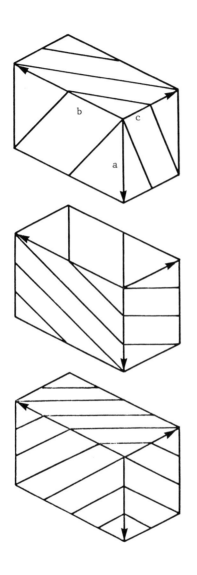

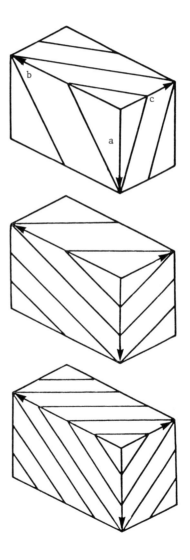

Stick to the axes shown on the first unit cell,

and index all the planes.

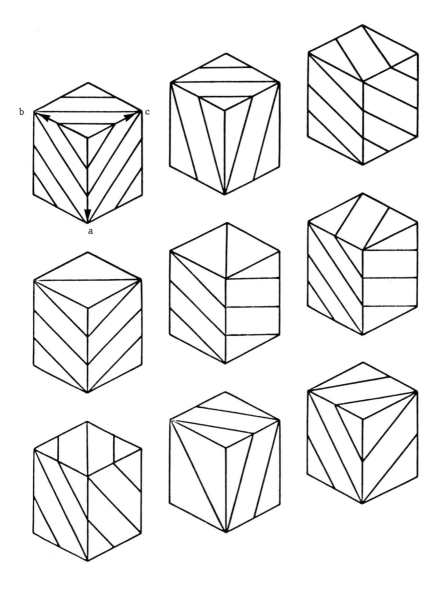

For a little variety at this stage, one set of axes
is used down the left hand side and another
down the right.

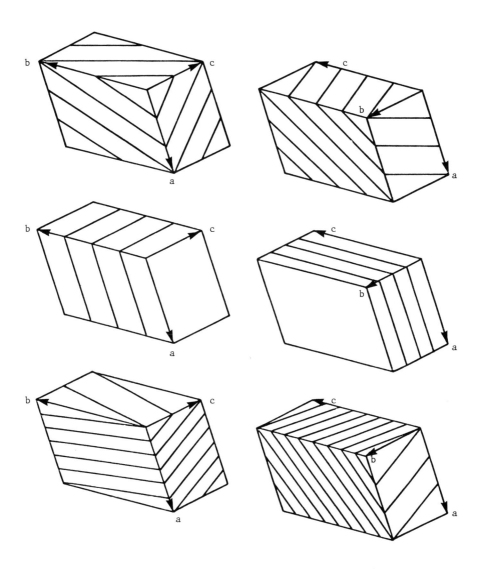

34

Index this last selection.

Take the a-axis as being roughly down
the page, the b-axis pointing
left and the c-axis right.

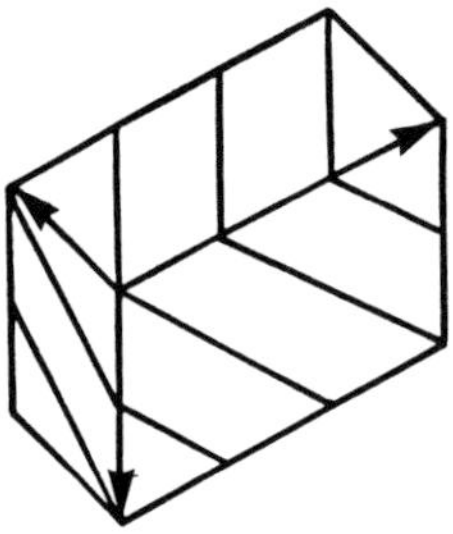

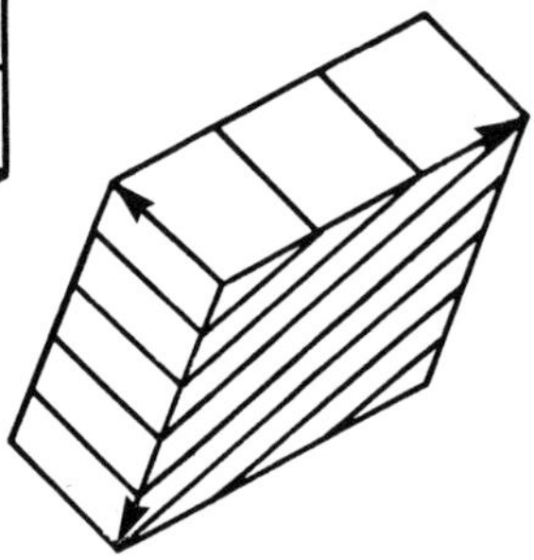

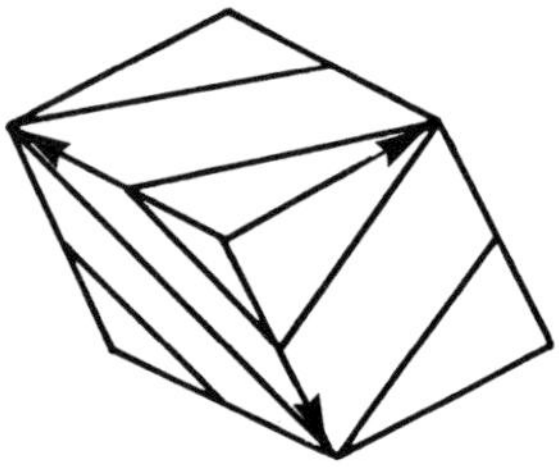

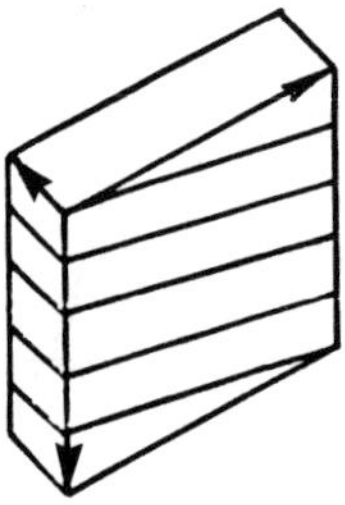

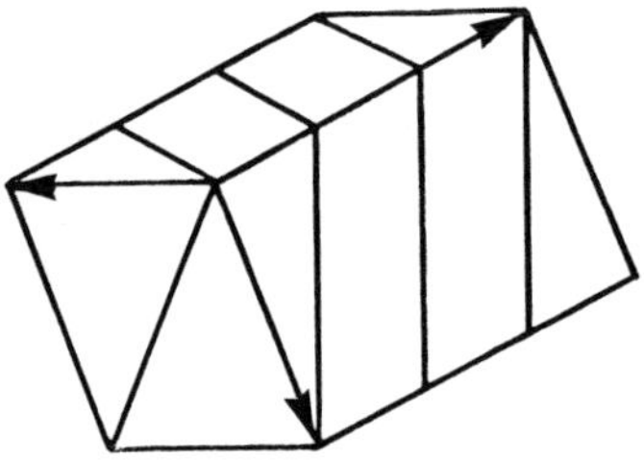

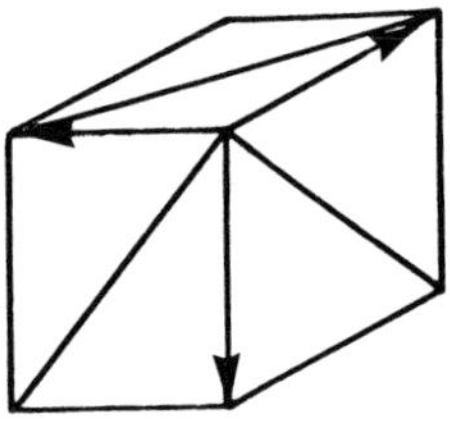

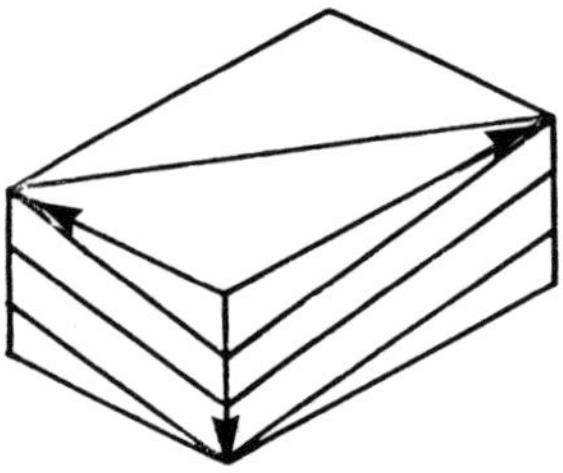

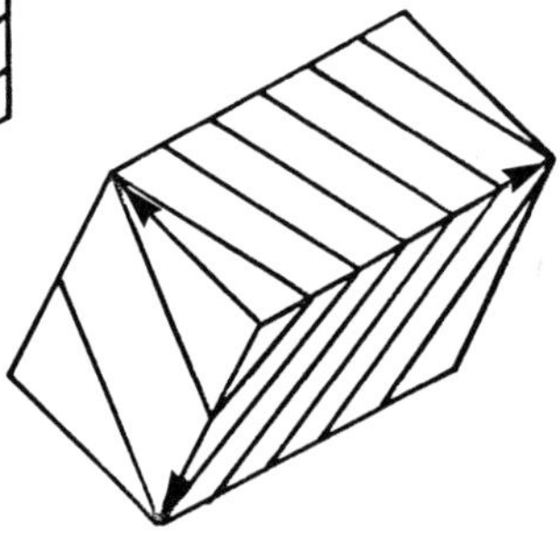

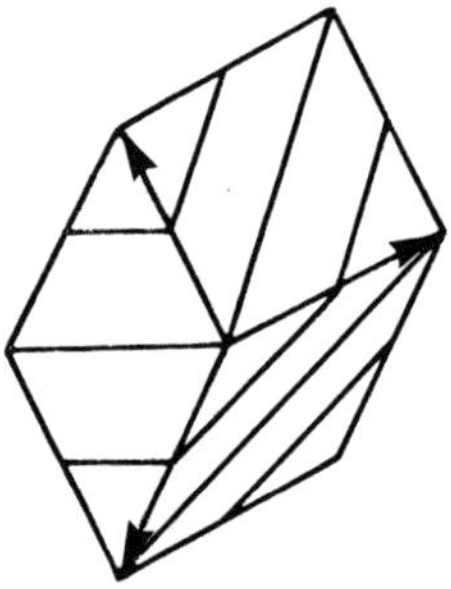

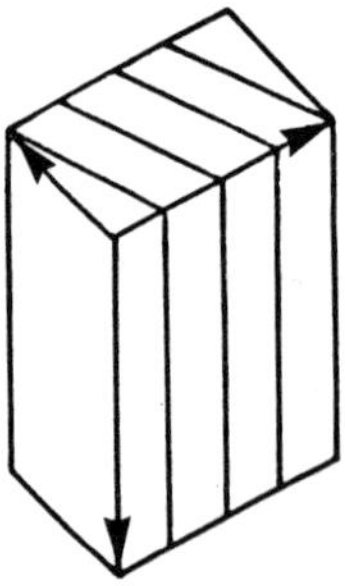

It must surely be clear by now.

Try sketching sets of planes in the unit cells below;
the edges are marked out for the purpose.   It's a tricky
job, particularly with negative indices, and usually
requires two or three attempts.

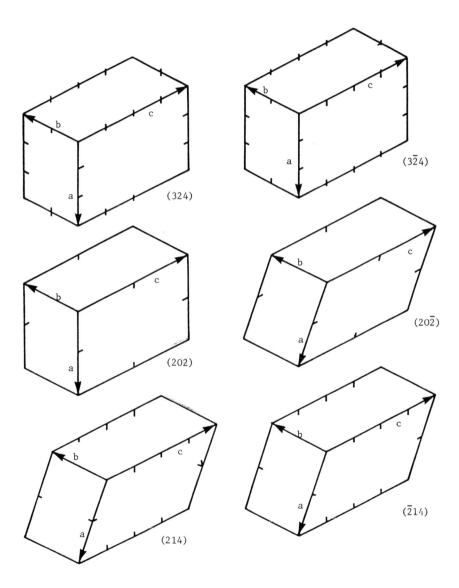

Miller-Bravais indices in the hexagonal system

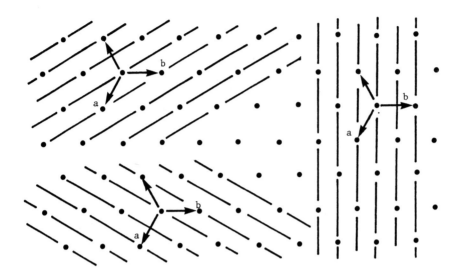

This is a hexagonal lattice with $\underline{c}$ pointing out of the paper.
We have also three sets of planes, all perpendicular to the paper, which
are clearly very similar to each other.

But their indices are (110), ($1\bar{2}0$) and ($\bar{2}10$), and their similarity is not
at all obvious in their indices.

So we introduce the third axis in the plane of the paper, and obtain the
extra index i in the same way as h, k and ℓ.   The indices are now

h	k	i	ℓ
1	1	$\bar{2}$	0
1	$\bar{2}$	1	0
$\bar{2}$	1	1	0

and they are seen to be related.

Notice that h + k + i = 0 every time.

The index i is really superfluous since, if h and k are given, the value of i
is always −(h + k).   It is often replaced by a dot   (hk.ℓ)

These 4-figure indices are used only in the hexagonal system and do not apply
to any other.

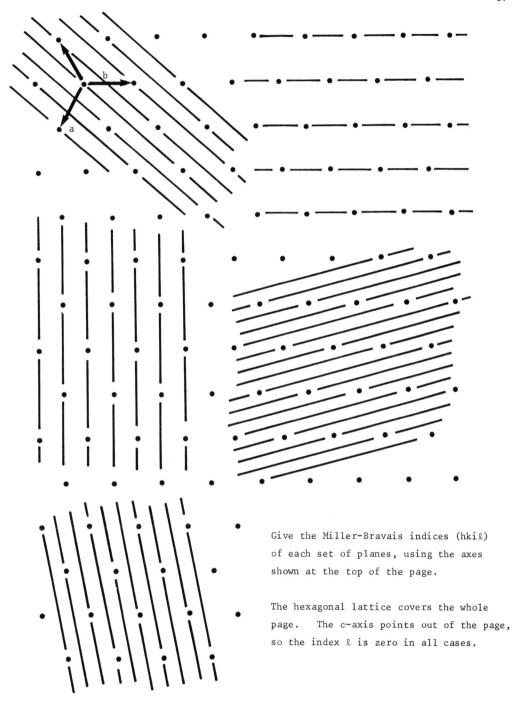

Give the Miller-Bravais indices (hkiℓ)
of each set of planes, using the axes
shown at the top of the page.

The hexagonal lattice covers the whole
page. The c-axis points out of the page,
so the index ℓ is zero in all cases.

Related sets of planes

(hkℓ) with its round brackets means a set of parallel planes.
{hkℓ} with curly brackets means all the sets of planes (hkℓ) which
     are related by symmetry.

In the cubic system,

        {100} means the 3 sets   (100) (010) (001)

        {110} means the 6 sets   (110) (101) (011)
                                                              ($\bar{1}$10) ($\bar{1}$01) (0$\bar{1}$1)

        {111} means the 4 sets   (111) ($\bar{1}$11) (1$\bar{1}$1) (11$\bar{1}$)

The advantage of this shorthand becomes very clear when you realise
that {123} means the 24 sets

    (1 2 3)   ($\bar{1}$ 2 3)   (1 $\bar{2}$ 3)   (1 2 $\bar{3}$)

    (1 3 2)   ($\bar{1}$ 3 2)   (1 $\bar{3}$ 2)   (1 3 $\bar{2}$)

    (2 1 3)   ($\bar{2}$ 1 3)   (2 $\bar{1}$ 3)   (2 1 $\bar{3}$)

    (2 3 1)   ($\bar{2}$ 3 1)   (2 $\bar{3}$ 1)   (2 3 $\bar{1}$)

    (3 1 2)   ($\bar{3}$ 1 2)   (3 $\bar{1}$ 2)   (3 1 $\bar{2}$)

    (3 2 1)   ($\bar{3}$ 2 1)   (3 $\bar{2}$ 1)   (3 2 $\bar{1}$)

(Notice that (1 $\bar{2}$ $\bar{3}$) for example has not appeared because it is
identical to ($\bar{1}$ 2 3))

All the examples on this page apply only to the cubic system with its three
identical axes.   Examples from other systems would not be quite so
straightforward, but it is highly unlikely that you will ever meet them.

Now write out the 24 sets of planes   {235}
               the 12 sets of planes   {122}
               the 12 sets of planes   {120}
               the 6 sets of planes   {220}
       and the 3 sets of planes   {200}

Directions in a Lattice

We need to be able to specify the direction through a lattice along which, say, an electron beam is passing, or along which the planes of atoms are sliding when the crystal is being deformed.

Each direction has indices [uvw] with square brackets.

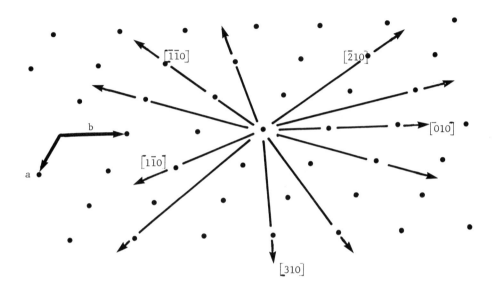

Study the diagram, and give the indices [uvw] of all the unindexed directions.

Assigning [uvw] to a direction is quite a different process from assigning (hkℓ) to a set of planes, and is much easier to understand.

Take [310] as an example. It means "For every three a-steps, take one b-step and no c-steps". By a-step we mean the distance from one point to the next in the a-direction, and of course we must be clear which axis is which before we start; if the axes are changed, [uvw] change too.

You will remember that the planes (khℓ) and ($\overline{hkℓ}$) were identical. Are the directions [uvw] and [$\overline{uvw}$] identical?

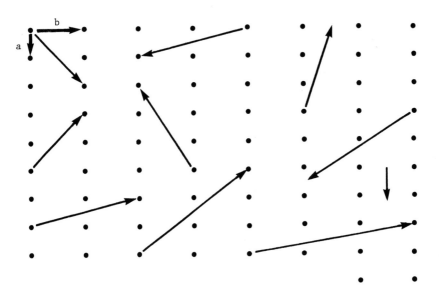

Give the indices [uvw] of all the directions shown in the orthorhombic
lattice above and the monoclinic lattice below.   The method is the
same, whatever the lattice, but be sure which axis is which before starting.

For simplicity here, all directions shown are in the plane of the paper,
so w is zero in all cases.

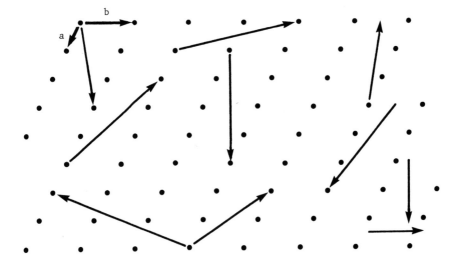

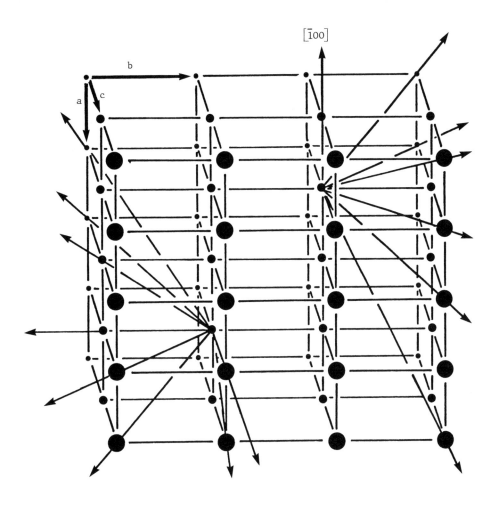

Give the indices [uvw] of each of the fourteen directions which are shown here radiating from two points in this orthorhombic P lattice.

Before starting, establish clearly in your mind the directions of a, b and c which are given in the far top left corner. The c-axis is intended to appear to be pointing towards you.

Related Directions < uvw>

$[uvw]$ with its square brackets indicates one direction
through the lattice.

< uvw> with pointed brackets indicates $[uvw]$ and all the similar
directions which are related to it by symmetry.

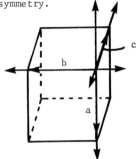

In the cubic system (only),

< 100> means the six directions

$$[100] \quad [\bar{1}00]$$
$$[010] \quad [0\bar{1}0]$$
$$[001] \quad [00\bar{1}]$$

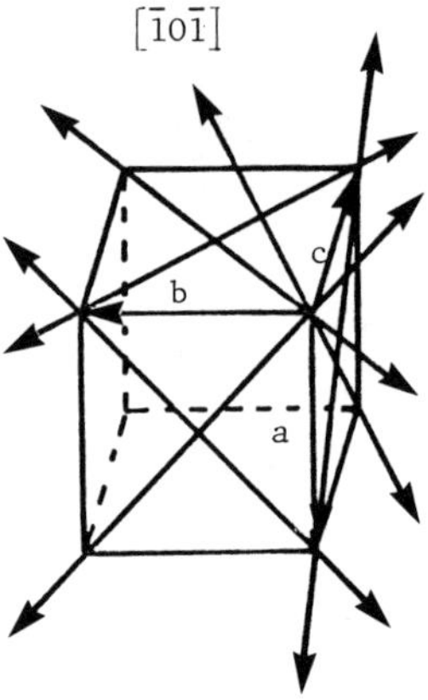

$[\bar{1}0\bar{1}]$

< 110> means the twelve directions

$$[110] \quad [\bar{1}\bar{1}0] \quad [1\bar{1}0] \quad [\bar{1}10]$$
$$[011] \quad [0\bar{1}\bar{1}] \quad [01\bar{1}] \quad [0\bar{1}1]$$
$$[101] \quad [\bar{1}0\bar{1}] \quad [10\bar{1}] \quad [\bar{1}01]$$

They are shown in the diagram, and
eleven of them need indices allocating.

Give, for the cubic system, all the directions which are covered
by each of the following.

< 111>   < 112>   < 123>

All the examples above are from the cubic system, since this is where
the nomenclature is most commonly encountered.

(This page is not important; it can be skipped).

Directions in a hexagonal lattice

We can use indices $[uvw]$ to describe these directions, as in any other lattice, or we can use what are known as Weber indices $[UVJW]$ which employ all four hexagonal axes and make $U + V + J = 0$. We did something similar with Miller indices (hkℓ) and Miller-Bravais indices (hkiℓ) for planes.

BUT IT IS MUCH TRICKIER WITH DIRECTIONS THAN WITH PLANES

In the diagram below, check that $[100]$ is also $[2\bar{1}\bar{1}0]$ and that $[210]$ is also $[10\bar{1}0]$

Notice that there is no obvious connection between the 3-index and the 4-index notations. The third index is NOT superfluous; the direction $[120]$ is NOT $[12\bar{3}0]$ but is in fact $[01\bar{1}0]$.

Mark on the diagram the directions $[2\bar{1}\bar{1}0]$, $[\bar{1}\bar{1}20]$, $[\bar{1}2\bar{1}0]$, $[01\bar{1}0]$, $[\bar{1}0\bar{1}0]$.

Now try giving the Weber indices of the other directions shown; you may find it difficult. Fortunately, you will only meet the Weber indices of very simple directions.

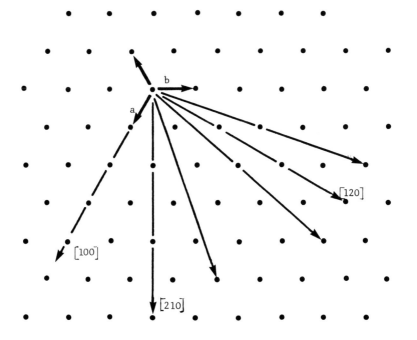

Vectors in a cubic lattice

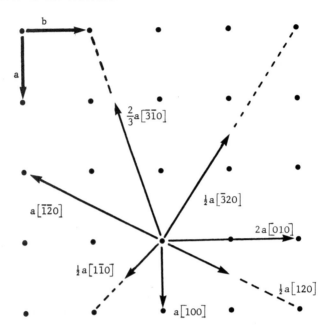

A shift (a translation) of a given distance in a lattice is
represented by a vector, and seven of these vectors are shown above.
All of them lie in the plane of the paper, so the index w is zero
in every case.

Study the symbol printed against each vector, and then try to give
the correct designation of each of the vectors in the diagram on the
next page.   The indices in brackets give the vector's direction in
the usual way, and the prefix is used to show the length of the vector.

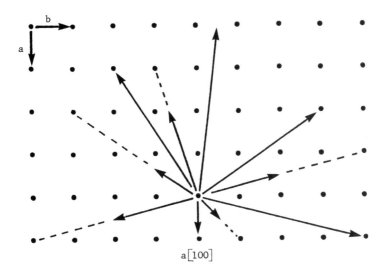

a[100]

The position in which the vector is drawn is of no importance, and the five vectors at the top of the diagram below are all $\frac{a}{3}$[130]

You may well find it convenient, of course, to imagine each vector moved to a lattice point in order to estimate its length and direction.

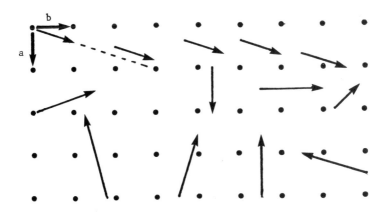

Addition of vectors

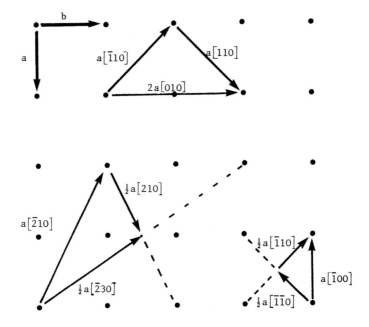

The three vector additions above may be written as

$$a[\bar{1}10] \quad + \quad a[110] \quad = \quad 2a[01\bar{0}]$$

$$a[\bar{2}10] \quad + \quad \tfrac{1}{2}a[210] \quad = \quad \tfrac{1}{2}a[\bar{2}30]$$

$$\tfrac{1}{2}a[\bar{1}\bar{1}0] \quad + \quad \tfrac{1}{2}a[\bar{1}10] \quad = \quad a[\bar{1}00]$$

Study these equations, and then try to give the equations corresponding to the vector additions shown on the next page.

The rule is simple. The total number of a-steps on the left hand side of the equation is equal to the number of a-steps on the right, and so on.

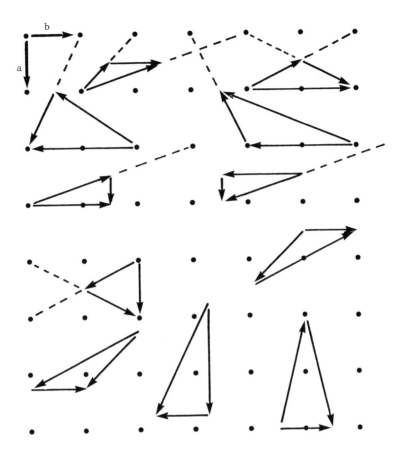

In each case, identify first of all the resultant in each of the
triangles.   It is the vector which is not nose-to-tail with
either of the others.

As always, be sure of the axes.   Changing the axes changes the
indices of vectors.

Now sketch, in the lattice below, the vector diagrams corresponding to

$$a[2\bar{1}0] \quad + \quad a[\bar{2}10] \quad = \quad 2a[010]$$

$$a[11\bar{0}] \quad + \quad a[\bar{1}10] \quad = \quad 2a[010]$$

$$a[110] \quad + \quad a[010] \quad = \quad a[120]$$

$$2a[010] \quad + \quad a[100] \quad = \quad a[120]$$

$$a[\bar{2}10] \quad + \quad a[110] \quad = \quad a[\bar{1}20]$$

Examine the relationships below, and state whether they are valid. In other words, state whether the two vectors on the left give the resultant on the right. It is not necessary to sketch them.

$$\frac{a}{2}[201] \quad + \quad \frac{a}{2}[0\bar{2}\bar{1}] \quad = \quad a[2\bar{1}0]$$

$$\frac{a}{6}[2\bar{1}\bar{1}] \quad + \quad \frac{a}{6}[11\bar{2}] \quad = \quad \frac{a}{2}[10\bar{1}]$$

$$\frac{a}{2}[321] \quad + \quad \frac{a}{3}[1\bar{1}\bar{2}] \quad = \quad \frac{a}{6}[11,4\bar{1}]$$

$$\frac{a}{6}[1\bar{2}1] \quad + \quad \frac{a}{6}[2\bar{1}\bar{1}] \quad = \quad \frac{a}{2}[1\bar{1}0]$$

$$\frac{a}{2}[10\bar{1}] \quad + \quad \frac{a}{2}[0\bar{1}1] \quad = \quad \frac{a}{2}[1\bar{1}0]$$

SECTION 3

Relations between planes and directions

This section presents simple methods of answering six questions which commonly arise in discussions of crystals. The first five are

> Does this direction lie in that plane?
> What is the direction along which these planes intersect?
> What plane is parallel to these two directions?
> Are these three directions coplanar?
> Do these three planes have a common line of intersection?

The methods we shall use apply to crystals of any of the seven systems.

The sixth question is more difficult, and it is frequently encountered in discussing the deformation of metals.

> What is the angle between these two directions
> (or between these two planes)?

The answer we shall give to this question applies only to the cubic system.

No proofs of any of these methods are given, and they are not necessary to someone who uses these methods as handy tools, but the origin of each method will be apparent to anyone familiar with the use of matrices and vectors.

DOES A DIRECTION $[uvw]$ LIE PARALLEL TO A SET OF PLANES $(hk\ell)$?

Yes, provided $hu + kv + \ell w = 0$

For example, $[11\bar{1}]$ lies parallel
to (213) because

$2.1 + 1.1 - 3.1 = 0$

Or again, $[\bar{1}3\bar{1}]$ lies parallel to (211)
because $-1.2 + 3.1 - 1.1 = 0$

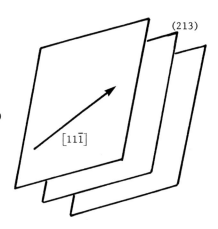

(213)

$[11\bar{1}]$

Does $[2\bar{1}\bar{1}]$ lie parallel to (111)?
" $[1\bar{1}0]$ " (111)?
" $[1\bar{2}2]$ " (41$\bar{1}$)?
" $[121]$ " (0$\bar{1}$2)?
" $[3\bar{1}\bar{1}]$ " (212)?
" $[22\bar{1}]$ " (1$\bar{1}$1)?
" $[132]$ " (1$\bar{1}$1)?
" $[111]$ " (2$\bar{1}$0)?

WHAT IS THE DIRECTION $[uvw]$ ALONG WHICH TWO SETS OF PLANES $(h_1\ k_1\ \ell_1)$ AND $(h_2\ k_2\ \ell_2)$ INTERSECT?

To answer this, set up the matrix

$$\begin{pmatrix} u & v & w \\ h_1 & k_1 & \ell_1 \\ h_2 & k_2 & \ell_2 \end{pmatrix}$$

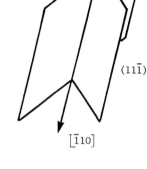

Then

$$u = k_1\ \ell_2 - k_2\ \ell_1$$
$$-v = h_1\ \ell_2 - h_2\ \ell_1$$
$$w = h_1\ k_2 - h_2\ k_1$$

For example, the intersection of (111) and (11$\bar{1}$)

$$\begin{pmatrix} u & v & w \\ 1 & 1 & 1 \\ 1 & 1 & \bar{1} \end{pmatrix} \quad \text{is} \quad [\bar{2}20] \quad \text{or, the same thing,}$$
$$[\bar{1}10] \quad \text{or} \quad [1\bar{1}0]$$

As the diagram shows, there are always two opposite directions which are equally valid answers to the question.

What are the directions of intersection of the following pairs of planes?

(100)	and	(010)	(111)	and	(100)
(110)	and	(1$\bar{1}$0)	(102)	and	($\bar{1}$20)
(11$\bar{1}$)	and	(111)	(303)	and	(535)
(211)	and	(010)	($\bar{2}$10)	and	(011)

GIVEN TWO DIRECTIONS $[u_1 \ v_1 \ w_1]$ and $[u_2 \ v_2 \ w_2]$, WHAT SET OF PLANES LIES PARALLEL TO BOTH OF THEM?

To put it another way, in what plane do

$[u_1 \ v_1 \ w_1]$ and

$[u_2 \ v_2 \ w_2]$ both lie?

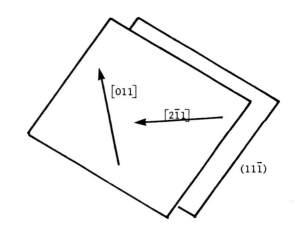

$(11\bar{1})$

Set up the matrix

$$\begin{pmatrix} h & k & \ell \\ u_1 & v_1 & w_1 \\ u_2 & v_2 & w_2 \end{pmatrix}$$

Then $h = v_1 w_2 - v_2 w_1$

$-k = u_1 w_2 - u_2 w_1$

$\ell = u_1 v_2 - u_2 v_1$

For example, in what plane do $[011]$ and $[2\bar{1}1]$ lie?

$$\begin{pmatrix} h & k & \ell \\ 0 & 1 & 1 \\ 2 & \bar{1} & 1 \end{pmatrix}$$

$h = 2$

$-k = -2$

$\ell = -2$

So they lie in

$(22\bar{2})$ or $(11\bar{1})$

Find the planes in which the following pairs of directions lie.

$[21\bar{2}]$ and $[1\bar{1}1]$ $[31\bar{0}]$ and $[00\bar{1}]$

$[10\bar{0}]$ and $[111]$ $[0\bar{1}0]$ and $[\bar{1}00]$

$[20\bar{1}]$ and $[1\bar{2}0]$ $[132]$ and $[21\bar{2}]$

$[\bar{1}02]$ and $[01\bar{2}]$ $[40\bar{1}]$ and $[110]$

DO THREE GIVEN DIRECTIONS $\begin{bmatrix} u_1 & v_1 & w_1 \end{bmatrix}$, $\begin{bmatrix} u_2 & v_2 & w_2 \end{bmatrix}$, $\begin{bmatrix} u_3 & v_3 & w_3 \end{bmatrix}$ LIE IN ONE PLANE?

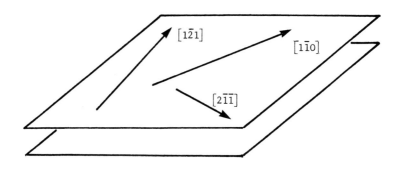

They do if the determinant

$$\begin{vmatrix} u_1 & v_1 & w_1 \\ u_2 & v_2 & w_2 \\ u_3 & v_3 & w_3 \end{vmatrix} \qquad \text{is zero}$$

For example, the directions $\begin{bmatrix} 1 & \bar{1} & 0 \end{bmatrix}$, $\begin{bmatrix} 1 & \bar{2} & 1 \end{bmatrix}$ and $\begin{bmatrix} 2 & \bar{1} & \bar{1} \end{bmatrix}$ are coplanar because

$$\begin{vmatrix} 1 & -1 & 0 \\ 1 & -2 & 1 \\ 2 & -1 & -1 \end{vmatrix} \;=\; 0$$

Do the directions in each of the following trios lie in one plane?

$\begin{bmatrix} \bar{2} & 2 & 0 \end{bmatrix}$	$\begin{bmatrix} 1 & \bar{2} & 1 \end{bmatrix}$	$\begin{bmatrix} 2 & \bar{1} & \bar{1} \end{bmatrix}$	$\begin{bmatrix} 2 & 1 & 2 \end{bmatrix}$	$\begin{bmatrix} 1 & 2 & 2 \end{bmatrix}$	$\begin{bmatrix} 3 & 3 & \bar{4} \end{bmatrix}$
$\begin{bmatrix} 3 & 2 & \bar{1} \end{bmatrix}$	$\begin{bmatrix} 1 & 4 & 2 \end{bmatrix}$	$\begin{bmatrix} 2 & \bar{2} & \bar{1} \end{bmatrix}$	$\begin{bmatrix} \bar{1} & 1 & 0 \end{bmatrix}$	$\begin{bmatrix} 1 & \bar{2} & 2 \end{bmatrix}$	$\begin{bmatrix} 2 & \bar{1} & \bar{2} \end{bmatrix}$
$\begin{bmatrix} 0 & 1 & 2 \end{bmatrix}$	$\begin{bmatrix} 1 & 1 & 0 \end{bmatrix}$	$\begin{bmatrix} 2 & 1 & 3 \end{bmatrix}$	$\begin{bmatrix} \bar{1} & 2 & \bar{1} \end{bmatrix}$	$\begin{bmatrix} 0 & 1 & 2 \end{bmatrix}$	$\begin{bmatrix} 1 & \bar{2} & 1 \end{bmatrix}$
$\begin{bmatrix} 0 & 1 & 0 \end{bmatrix}$	$\begin{bmatrix} 1 & 2 & 3 \end{bmatrix}$	$\begin{bmatrix} 2 & 3 & 6 \end{bmatrix}$	$\begin{bmatrix} 4 & 1 & 0 \end{bmatrix}$	$\begin{bmatrix} 1 & 0 & 0 \end{bmatrix}$	$\begin{bmatrix} 3 & 1 & 0 \end{bmatrix}$

A group of planes with a common line of intersection is called a "zone" of planes.   The line of intersection is the "zone axis".

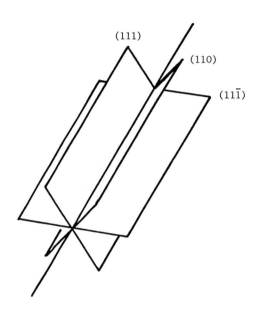

(111)

(110)

(11$\bar{1}$)

THREE SETS OF PLANES $(h_1 \ k_1 \ \ell_1)$
$(h_2 \ k_2 \ \ell_2)$ AND $(h_3 \ k_3 \ \ell_3)$ BELONG
TO ONE ZONE IF THE DETERMINANT

$$\begin{vmatrix} h_1 & k_1 & \ell_1 \\ h_2 & k_2 & \ell_2 \\ h_3 & k_3 & \ell_3 \end{vmatrix} \quad \text{IS ZERO}$$

The planes (111), (110) and (11$\bar{1}$)

belong to one zone because

$$\begin{vmatrix} 1 & 1 & 1 \\ 1 & 1 & 0 \\ 1 & 1 & -1 \end{vmatrix} = 0$$

Examine each of the following trios of planes, and determine whether its members belong to a zone.

(2$\bar{1}\bar{1}$)	(233)	(111)	(122)	(212)	(433)
(212)	(236)	(1$\bar{1}\bar{2}$)	(103)	(001)	(104)
(302)	(111)	(210)	(201)	($\bar{2}$0$\bar{3}$)	(120)
(001)	(312)	(623)	(102)	($\bar{1}$2$\bar{2}$)	(122)

WHAT IS THE ANGLE BETWEEN TWO DIRECTIONS $[u_1 \; v_1 \; w_1]$ and $[u_2 \; v_2 \; w_2]$ IN A CUBIC LATTICE?

If the angle is ϕ,

$$\cos \phi = \frac{u_1 u_2 + v_1 v_2 + w_1 w_2}{\sqrt{u_1^2 + v_1^2 + w_1^2} \; \sqrt{u_2^2 + v_2^2 + w_2^2}}$$

For example, between $[101]$ and $[01\bar{1}]$ the angle is given by

$$\cos \phi = \frac{0 + 0 - 1}{\sqrt{2} \quad \sqrt{2}} = -\frac{1}{2}$$

So $\phi = 120^\circ$

It is fairly obvious that a direction such as $[101]$ in a cubic lattice is perpendicular to the planes (101), so the formula used above to find the angle between $[101]$ and $[01\bar{1}]$ applies equally to finding the angle between the planes (101) and $(01\bar{1})$.

Remember that everything on this page refers to the cubic system only. It is not true for any other.

Calculate the angles between the following pairs of directions

$[111]$	$[3\bar{2}\bar{1}]$	$[211]$	$[\bar{1}\bar{2}1]$
$[110]$	$[011]$	$[110]$	$[211]$
$[3\bar{1}0]$	$[\bar{1}31]$	$[210]$	$[1\bar{2}1]$
$[\bar{1}00]$	$[110]$	$[\bar{1}\bar{1}0]$	$[212]$

SECTION 4

Atomic coordinates

To deal with the planes and directions in the crystal we have
had to think of the crystal lattice, rather than the real
material of the structure, but now we return to the atoms
themselves and place them in the unit cell to rebuild the
structures. We do this by specifying each atom's position
by its fractional coordinates, and the use of these is
explained by reference to a few structures which are commonly
encountered.

A crystal structure is described by stating the "fractional coordinates" xyz of each atom in the unit cell.

The structure of copper is usually shown in this way, with atoms at the cube corners and the face centres. In the diagram the atoms on the back faces are drawn with broken lines.

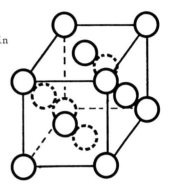

Its lattice is cubic F, and clearly each lattice point represents one copper atom.

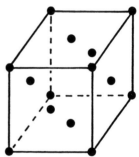

Notice that there are only four atoms (and four lattice points). If the four shown here are repeated in every unit cell, the complete structure is formed.

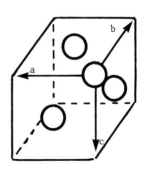

The fractional coordinates of the four atoms are

$$000 \quad \tfrac{1}{2}\tfrac{1}{2}0 \quad 0\tfrac{1}{2}\tfrac{1}{2} \quad \tfrac{1}{2}0\tfrac{1}{2}$$

The crystal structure of iron at room
temperature is shown, with an atom at
each cube corner and one at the
body centre.

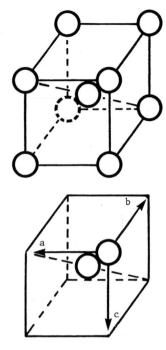

Its lattice, not drawn here, is cubic I,
with one lattice point in place of each
of these atoms.

To each cube there are only two atoms
(and two lattice points).   Their
fractional coordinates are

000   (at the origin)
and $\frac{1}{2}\frac{1}{2}\frac{1}{2}$ (at the body centre)

Don't be misled by the upper diagram
into thinking that there are nine atoms
in the cube.

Magnesium has the hexagonal structure on the left below, and its lattice,
the hexagonal P lattice, is not centred in any way.   Each unit cell
contains only one lattice point, but two atoms.
The fractional coordinates of the two atoms are

0 0 0 (at the origin)
and $\frac{2}{3} \frac{1}{3} \frac{1}{2}$ (inside the unit cell and half way up)

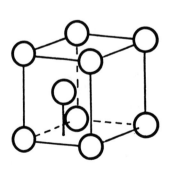

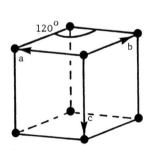

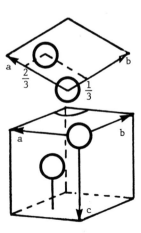

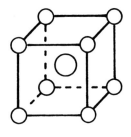 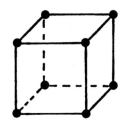 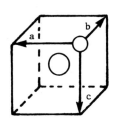

On the left is the structure of CsCl, with the Cs ions at the corners
and the slightly larger Cl ion at the middle of the cube.
At each lattice point there is one pair of ions, as shown on the
right.   If a pair like this is placed at every lattice point, the
whole structure is built up.

Name its lattice, which is shown in the centre.
Give the fractional coordinates of the Cs ion and the Cl ion.

(Don't imagine that CsCl has a body-centred cubic structure.
There is no such thing.   Only <u>lattices</u> are bcc, not structures, and
CsCl certainly doesn't have one.)

In $CaTiO_3$, the large Ca is at the cube centre and the small Ti at the
cube corners.   Its lattice, as you can see, is the same as that of
CsCl above.   There are five atoms to each lattice point.

Name the lattice.
Give the fractional coordinates of the Ca, the Ti and each of the
three oxygens.

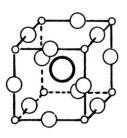 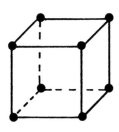 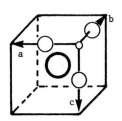

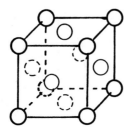 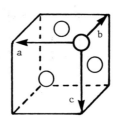

This is Ni$_3$Al, an important structure in some nickel-based alloys.
The Al (bold circles) are at the cube corners.

State the lattice type.   (It is not cubic F, though pure nickel
has the cubic F lattice.)

Give the fractional coordinates of the four atoms in the unit cell.

Two of the diagrams below show the cubic structure of ReO$_3$, with different
positions chosen for the origin of the unit cell.   One of them shows
more clearly than the other the octahedral coordination of O around the Re.

State the fractional coordinates of the (four) atoms in each of the
unit cells.

Name the lattice type.

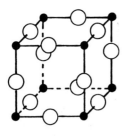 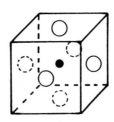 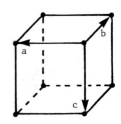

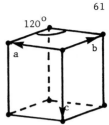

The three structures below all have the
hexagonal P lattice, with one lattice point
per unit cell.

For all three, state the fractional coordinates
of the atoms.   (Don't, of course, give the
coordinates of every atom in the diagram).

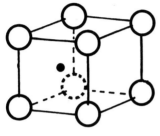

Tungsten carbide, WC, with the large W
at the unit cell corners.

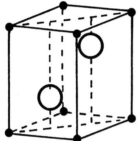

Another tungsten carbide, W_2C, shown
with the large W at $\frac{1}{4}$ and $\frac{3}{4}$ of the
way up the unit cell.

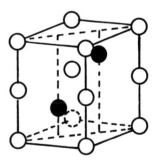

Nickel arsenide, NiAs, with
Ni at the corners and on the
cell edges, and the black As
placed symmetrically between
the layers of Ni.

Notice how to read the fractional
coordinates x and y.

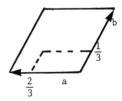

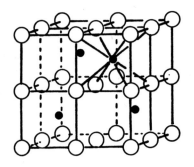 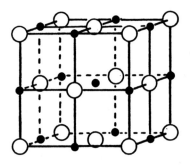

CaF_2 is on the left above and NaCl on the right.   The large ions
are F and Cl.   To show the structures more clearly, each unit cell
has been subdivided into eight.

Both structures have the same cubic lattice.   Which of the three
possible ones (P, F or I) is it?

How many atoms are there to a lattice point in each of the structures?

State their fractional coordinates.

The answers are partly provided by the sketches below.   If the group of
three or two atoms is placed at each point of the correct lattice, the
structures at the top of the page will be produced.

Notice that simply by stating the lattice type and the fractional coordinates
of only three or two atoms the entire structure is described.

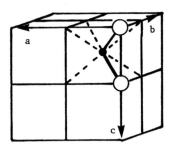 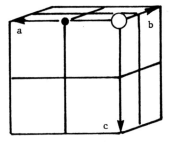

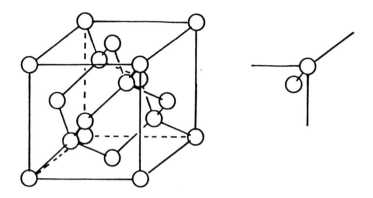

The two very similar structures, diamond above and zincblende, ZnS, below, have the same cubic lattice.

In both cases, if the fractional coordinates of only two atoms (shown separately) are given, together with the lattice type, the entire structure is described.

Name the lattice type.

Give the fractional coordinates of all eight atoms in each unit cell (not just the two atoms shown).

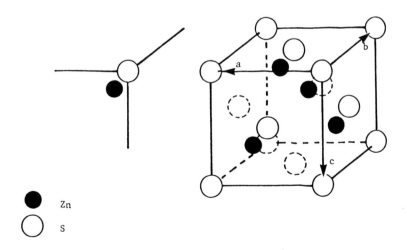

●   Zn

○   S

64

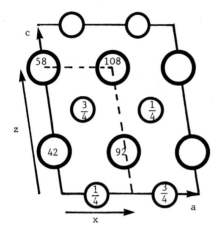

CuO is monoclinic. Notice how the x and z fractional coordinates of an atom are measured when the axes are not orthogonal.

There are four copper and four oxygen atoms with the coordinates given below. Identify each of them on the diagram.

Cu (small circles) $\frac{1}{4}$ $\frac{1}{4}$ 0 $\frac{3}{4}$ $\frac{1}{4}$ $\frac{1}{2}$

$\frac{3}{4}$ $\frac{3}{4}$ 0 $\frac{1}{4}$ $\frac{3}{4}$ $\frac{1}{2}$

O (large circles) 0 0.42 $\frac{1}{4}$ $\frac{1}{2}$ 0.92 $\frac{1}{4}$

0 0.58 $\frac{3}{4}$ $\frac{1}{2}$ 0.08 $\frac{3}{4}$

To help in visualising the structure, the atoms are marked with their y-coordinates. In some cases this is given as a fraction, and in others it is given in hundredths, so that 42 means 42/100.

One oxygen is 108/100 above the paper, and it could have been shown as 8/100, since 108/100 is in the next unit cell, but this atom has been chosen to show how it is packed near the others in the diagram.

SECTION 5
X-ray powder photographs of cubic crystals

Almost the whole of our knowledge of the structure of crystals,
the size and shape of the unit cells and the positions of the
atoms within them, has been obtained by observing the
diffraction of x-rays by the crystals. To determine a
crystal structure is a large undertaking, but one piece of
work which is well within the scope of a student in say the
first term of a university course is to discover whether a
crystal is cubic, to identify its lattice (P, F or I) and to
measure the lattice parameter. That is the material of this
section.

The technique employs the x-ray powder camera. It is a
beautifully simple instrument to operate, but its diffraction
patterns can be baffling at first acquaintance. However, once
two or three have been successfully interpreted (and fourteen
are illustrated here) they present little difficulty, and an
ability to interpret these patterns is a powerful skill for a
student to acquire.

Diffraction and reflection of x-rays

The section opens with a brief reminder of the Bragg law which
all students of this subject will have encountered, and it concerns
the "reflection" of x-rays by planes in a crystal. Yet we speak
also of x-ray "diffraction" and of diffraction patterns. The
two terms describe two different conceptual approaches to the
same phenomenon, and we can choose which to adopt, but in x-ray
powder work the reflection approach is much the more direct and
workmanlike. All our beams will be "reflected" by the crystal,
Bragg-fashion. Nevertheless, when they strike a photographic
film they give a diffraction pattern; we just never speak of a
reflection pattern.

Reflection of x-rays

An x-ray beam is reflected, faintly,
by a layer of atoms, just as
light is reflected
by a mirror

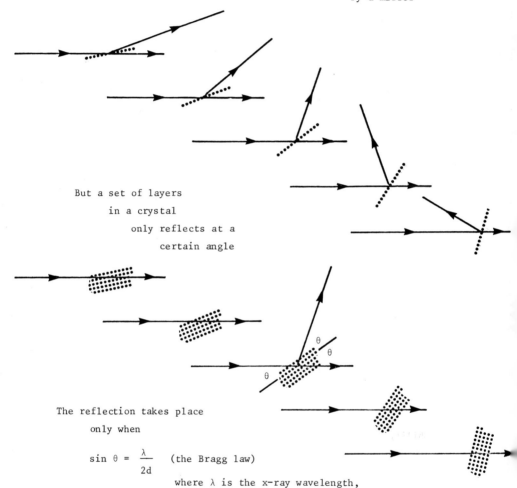

But a set of layers
in a crystal
only reflects at a
certain angle

The reflection takes place
only when

$$\sin \theta = \frac{\lambda}{2d} \quad \text{(the Bragg law)}$$

where λ is the x-ray wavelength,
and d is the spacing between the planes in the crystal.

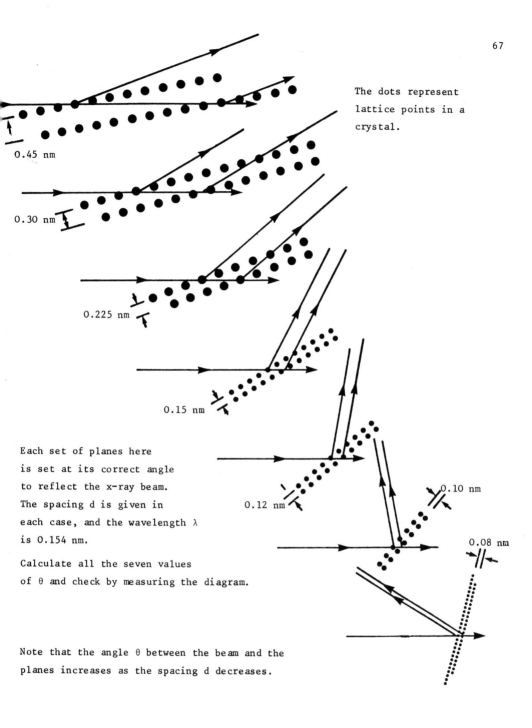

The dots represent lattice points in a crystal.

0.45 nm

0.30 nm

0.225 nm

0.15 nm

Each set of planes here
is set at its correct angle
to reflect the x-ray beam.
The spacing d is given in
each case, and the wavelength λ
is 0.154 nm.

Calculate all the seven values
of θ and check by measuring the diagram.

0.12 nm

0.10 nm

0.08 nm

Note that the angle θ between the beam and the
planes increases as the spacing d decreases.

68

Spacings of planes in a cubic crystal

On the previous page the planes were selected from seven different crystals. But planes of many spacings exist within the same crystal, and their spacings depend on a and on hkℓ

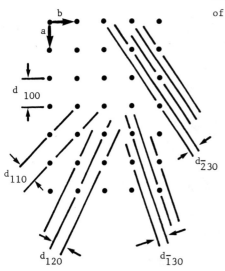

The spacing $d_{hkℓ}$ of the planes is

$$d_{hkℓ} = \frac{a}{\sqrt{h^2 + k^2 + ℓ^2}}$$

Suppose the cubic unit cell has side a = 1.20 nm (the lattice parameter as it is called).

Complete the following (in your head, no calculator needed)

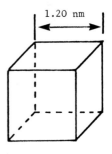

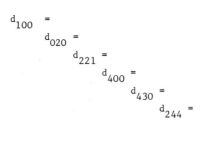

d_{100} =

d_{020} =

d_{221} =

d_{400} =

d_{430} =

d_{244} =

The x-ray powder camera (or Debye-Scherrer camera)

Brittle crystals are ground in a mortar to form a fine powder and
put into a glass capillary of about 0.5 mm dia. with thin walls which
allow the x-rays to penetrate.   Crystals of ductile metal will be
in the form of filings which must be annealed to remove the deformation
caused by the filing.

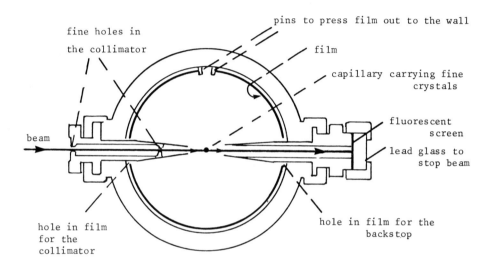

The capillary is at the centre of the light-tight metal cylinder, and
the film on which the reflected beams are recorded is pressed tightly
against the cylinder wall.   The circumference is exactly 180 mm.

X-rays enter through holes about 0.5 mm wide in the metal collimator,
and the unreflected beam is caught in a backstop.

The capillary is rotated by an electric motor to give all crystals the
chance of reflecting the beam.

Exposure times are from 15 minutes to several hours.

These diagrams show the position of the film in the camera and the cones of reflected beams radiating from the finely powdered specimen.

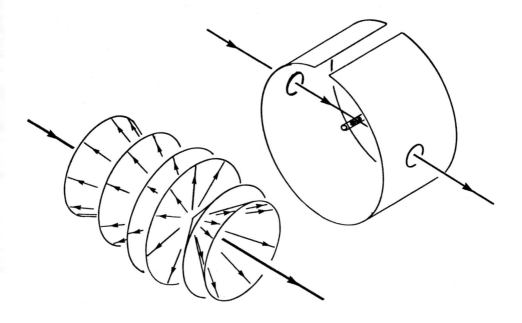

The cones of beams form arcs on the film, shown opened out below. If the crystals are coarsely ground, the arcs will appear speckled.

It appears strange at first sight that such precise cones of beams radiate from the randomly arranged crystals in the capillary, but each set of planes such as (100) has a certain spacing d_{100}, so its θ_{100} is fixed. Any crystal not at the correct angle θ will not reflect the beam, as is shown on the next page.

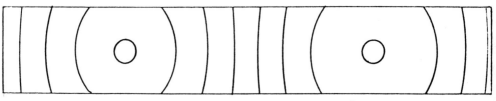

entrance hole   exit hole

The diagram shows fine crystals within their thin-walled glass capillary.

The {100} planes of some crystals are reflecting the x-rays at
$2\theta = 40°$, and the {110} planes of others are reflecting at
$2\theta = 58°$.

Most crystals are not reflecting at all, because they are not in the correct position.

Things appear very haphazard in the capillary, but beams are produced only at the precise angles $40°$ and $58°$, as the lower diagram shows.

(In practice, other beams would be produced at other angles, but only 100 and 110 are shown for simplicity).

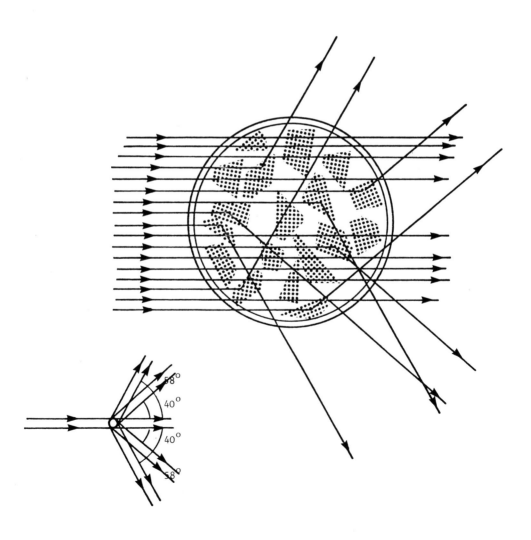

Calculating the lattice parameter

The specimen here is β-brass (CuZn). The angle $2\theta_{hk\ell}$ is shown by each reflected cone, and the arcs are on the film below. The x-ray wavelength $\lambda = 0.229$ nm.

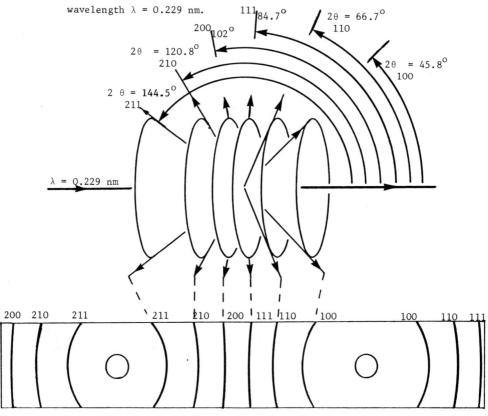

Calculate the lattice parameter a from each of the arcs.

hkℓ	2θ	$d_{hk\ell} = \lambda/2\sin\theta$	$a = d\sqrt{h^2+k^2+\ell^2}$	
100	45.8°	0.294 nm	$0.294\sqrt{1^2+0^2+0^2}$	= 0.294 nm
110	66.7°	0.208	$0.208\sqrt{1^2+1^2+0^2}$	= 0.295 nm
111	84.7°		$d\sqrt{1^2+1^2+1^2}$	= 0.294 nm
200	102°		$d\sqrt{2^2+0^2+0^2}$	=
210	120.8°		$d\sqrt{2^2+1^2+0^2}$	=
211	144.5°		$d\sqrt{2^2+1^2+1^2}$	=

The alloy BeCo is cubic
with a = 0.261 nm.
Check this using any of
the θ-values of the
five cones.

(N.B.   2θ is given,
and you need θ.)

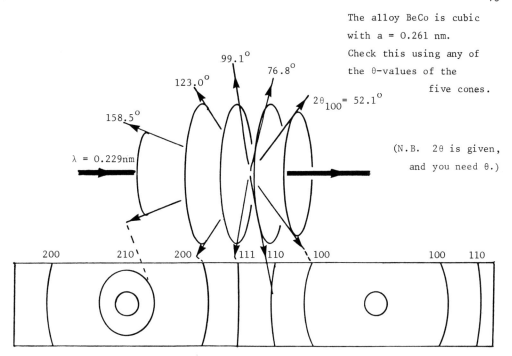

The alloy LiTl below has a larger unit cell and, with smaller
θ-values, gives seven cones.   Check that a = 0.342 nm.

Note the large gap between 211 and 220.

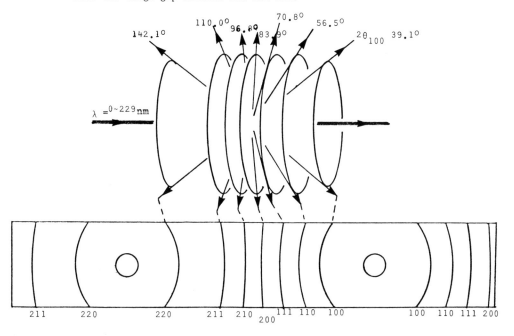

This pattern is given by MgTℓ (a = 0.3628nm) using CoKα radiation (λ= 0.1790nm). Examine the sequence of indices, and provide indices for the last six arcs.

Notice the gaps at

$\sqrt{7}$ and $\sqrt{15}$

(and all $\sqrt{8n-1}$)

There are always two wavelengths in the x-ray beam, which cause the highest arcs to be doubled (not shown on previous pages). They are

CoKα$_2$ 0.17928nm
and CoKα$_1$ 0.17889nm

and in the lower arcs they are not resolved.

hkℓ	$\sqrt{h^2+k^2+\ell^2}$
100	$\sqrt{1}$
110	$\sqrt{2}$
111	$\sqrt{3}$
200	$\sqrt{4}$
210	$\sqrt{5}$
211	$\sqrt{6}$
220	$\sqrt{8}$
221 & 300	$\sqrt{9}$
	$\sqrt{10}$
	$\sqrt{11}$
	$\sqrt{12}$
	$\sqrt{13}$
	$\sqrt{14}$
	$\sqrt{16}$

Every film in this book has its exit hole on the right. The holes on other films can be recognised by the doubled arcs around the entrance hole.

The distance between the points where the beam enters and leaves
the film is very close to 90mm, and because of this the distances in
mm shown between the arcs are also 2θ in degrees.

First index each of the arcs, and then take any of the measurements
of 2θ to obtain the lattice parameter.
(Use $d = \lambda/2\sin\theta$ and $a = d \sqrt{h^2+k^2+\ell^2}$).

Both patterns were taken with $\lambda= 0.194$ nm (FeKα).

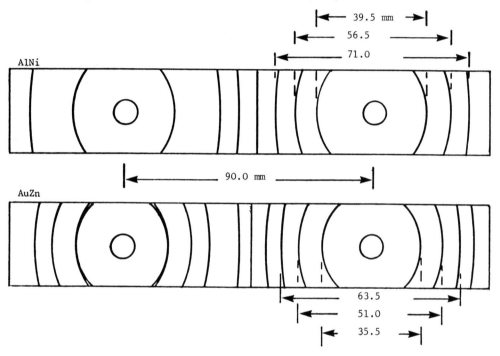

AlNi

|← 39.5 mm →|

|← 56.5 →|

|← 71.0 →|

|← 90.0 mm →|

AuZn

|← 63.5 →|

|← 51.0 →|

|← 35.5 →|

Measurements like these, simply taken with a ruler to the nearest 0.5mm ,
will give the lattice parameters of the two alloys correct to about ±1%.

To obtain an accurate value the arcs at the high-θ end of the film
must be measured.   There are several reasons for this, at least one of
them obvious, and the method is given on the next page.

To obtain an accurate lattice parameter
of this BeCo alloy, measure S and S_o.

Then $\dfrac{S}{S_o} = \dfrac{360 - 4\theta}{180}$

and $\theta = 90 \left(1 - \dfrac{S}{2S_o}\right)$

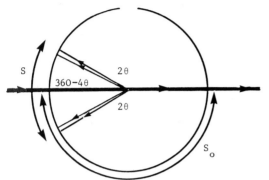

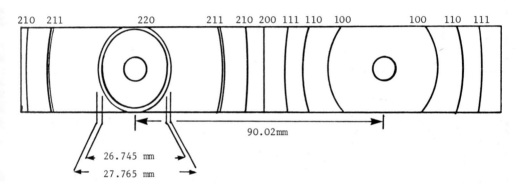

26.745 mm

27.765 mm

90.02mm

The x-rays are from a cobalt target, for which

λ of $CoK\alpha_2$ = 0.179278 nm

and λ of $CoK\alpha_1$ = 0.178892 nm

For the highest line on the film, $\theta = 90 \left(1 - \dfrac{26.745}{2 \times 90.02}\right)$

Then $a = \dfrac{0.179278}{2 \sin\theta} \cdot \sqrt{8} = 0.2606$ nm

For the other line of the doublet, $\theta = 90 \left(1 - \dfrac{27.765}{2.90.02}\right)$

Then $a = \dfrac{0.178892 \cdot \sqrt{8}}{2 \sin\theta} = 0.2606$ nm

Calculate the lattice parameters of these three alloys using the values of S and S_o shown (in mm).

All the films are taken with FeKα radiation, for which

$$\lambda \text{ of } K\alpha_2 = 0.193991 \text{ nm}$$

$$\lambda \text{ of } K\alpha_1 = 0.193597 \text{ nm}$$

and the mean $\lambda = 0.193728$ nm   (This is not the arithmetic mean because the α_1 beam is more intense than the α_2.)   For BeCu use the mean λ

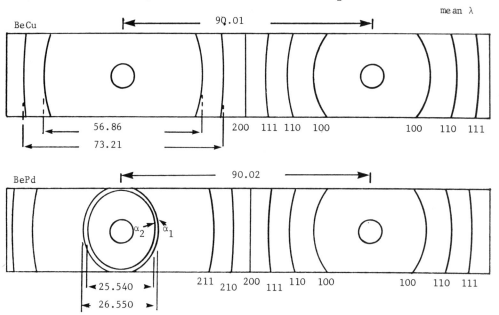

For BePd and LiAg take care to use the correct λ

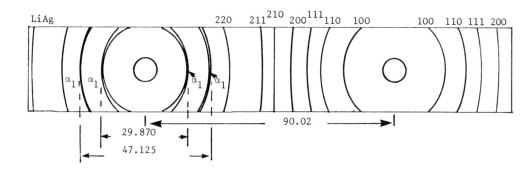

The f.c.c. pattern

This irregular pattern of arcs is given by any structure with the
face-centred cubic (cubic F) lattice, in this case LiF which has the
NaCl structure.   Anything with the cubic F lattice can be recognised
instantly by this pattern.   (All the patterns on earlier pages are of
alloys with the cubic P lattice).

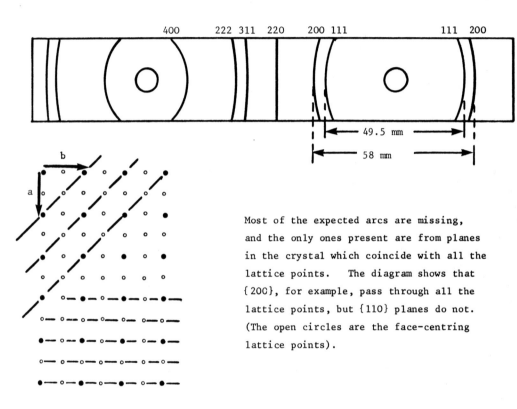

Most of the expected arcs are missing,
and the only ones present are from planes
in the crystal which coincide with all the
lattice points.   The diagram shows that
{200}, for example, pass through all the
lattice points, but {110} planes do not.
(The open circles are the face-centring
lattice points).

Notice on the film the general rule that
arcs are present for which h, k and ℓ are all odd or
all even.

Use the measurements on the 111 and 200 arcs to obtain the lattice
parameter of LiF.   The radiation used was FeKα, with λ = 0.194 nm.

The b.c.c. pattern

There appears to be no gap after the sixth arc on this pattern given by tungsten powder. Tungsten has the cubic I lattice (bcc) and only half the expected arcs are present.

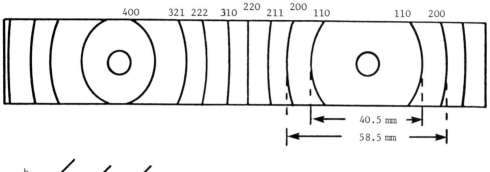

Notice the indices, which show that only the arcs $\sqrt{2}$, $\sqrt{4}$, $\sqrt{6}$ and so on are present, and the gap at $\sqrt{7}$ is camouflaged.

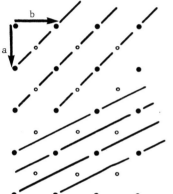

This time, {110} planes cut all the lattice points, but {210}, for example, do not and give no reflection. (The open circles are the body-centring lattice points).

The general rule for a bcc lattice is that <u>reflections are present</u> <u>when the sum h + k + ℓ is even</u>.

Use the pattern to calculate the approximate lattice parameter of tungsten. The x-radiation is CuKα, with $\lambda = 0.154$ nm.

These three patterns are given by, from the top, the metals copper and
tantalum, and the compound silver fluoride.   The wavelength used
is given on each film.

Identify the lattice in each case, simply by considering the arrangement
of arcs, and use the measurements given to calculate approximate values
of the lattice parameters.

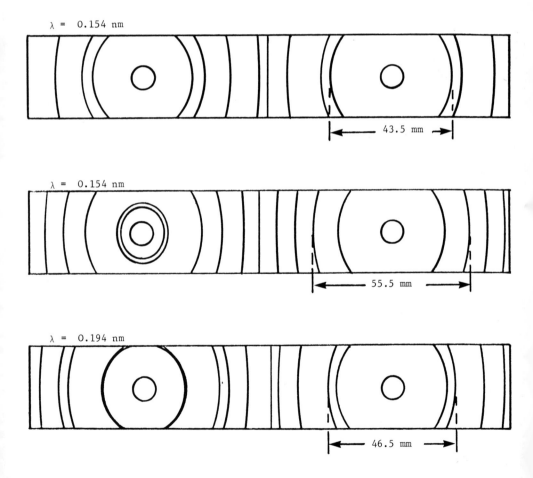

λ = 0.154 nm

|← 43.5 mm →|

λ = 0.154 nm

|← 55.5 mm →|

λ = 0.194 nm

|← 46.5 mm →|

SECTION 6

The reciprocal lattice and the diffraction
(or reflection) of waves by a crystal

It is difficult to imagine how a beam of x-rays entering a
crystal might behave, with so many planes set at all angles to
the beam and all with different d-spacings.   Which of them,
if any, will satisfy the Bragg law and reflect the beam?

We are going to show that the question can be answered very
simply by the use of a device known as the reciprocal lattice.
It is an abstract affair, existing in reciprocal space where
distances are measured in reciprocal nanometres, yet it is a
useful practical tool for dealing with problems of diffraction.
The more it becomes familiar, the more it becomes an indispensable
intellectual instrument, and metallurgists, physicists and
chemists who deal with diffraction, and particularly electron
microscopists as we shall see in Section 7, much prefer to
see diffraction taking place in reciprocal space rather than
in real space.

We make a start here in constructing the reciprocal lattice
(on the right) of a monoclinic real lattice (on the left), and
we go no further than setting up the reciprocal lattice vectors
$\underline{a}^*$ and $\underline{b}^*$

> $\underline{a}^*$ is perpendicular to the (100) planes
>
> $\underline{b}^*$ is   "   "   (010)   "

The length of $\underline{a}^*$ is $1/d_{100}$, and of $\underline{b}^*$, $1/d_{010}$

Direct (or real) lattice   Reciprocal lattice in
in real space   reciprocal space

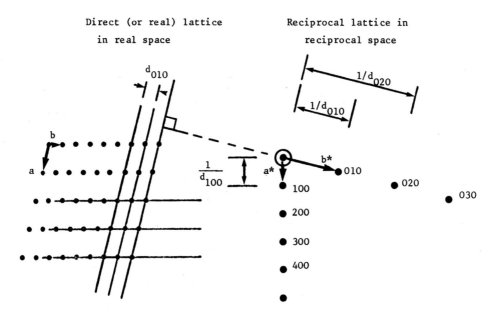

Each point such as 020 on the right represents the (020) planes on
the left, and it is $1/d_{020}$ from the origin.

More points have now been added, in a fairly obvious way, to build up more of the reciprocal lattice, and every point added represents a set of planes in the direct lattice.

As an example, notice what is called the $2\bar{1}0$ reciprocal lattice vector on the right and the $(2\bar{1}0)$ planes in the real lattice on the left. Check, by measuring, that they are at right angles and that the length of the vector is $1/d_{2\bar{1}0}$, just as that of $\underline{a}^*$ was $1/d_{100}$.

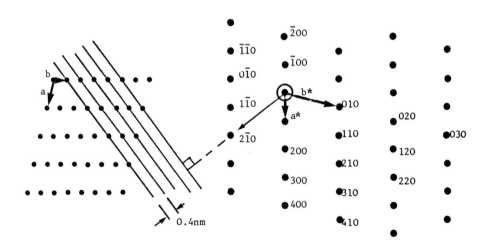

Real space   Reciprocal space

Think of the page in two parts, real space on the left and reciprocal space on the right. Any planes you care to draw in the real lattice will be perpendicular to their vector in reciprocal space.

Complete the indexing of the reciprocal lattice.

The next step is to set up $\underline{c}$* of the reciprocal lattice perpendicular to the (001) planes. Since $\underline{c}$ is short, 0.4nm, $\underline{c}$* is long, $2.5nm^{-1}$.

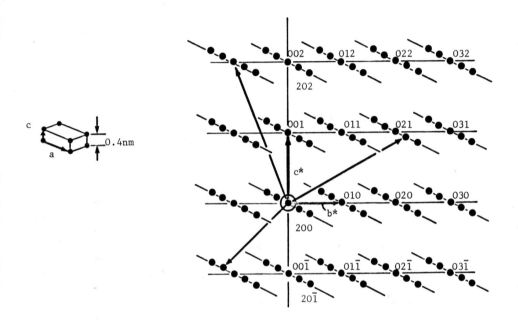

In this perspective sketch of the reciprocal lattice, the net with the origin is the hk0 net from the previous page, and the others are identical, but with their points representing the (hk2), (hk1) and (hk$\bar{1}$) planes.

Give the indices of the three reciprocal lattice vectors shown on the diagram.

This time we construct the reciprocal lattice of an orthorhombic
real lattice.   First, as before, we lay out vectors a* and b*
perpendicular to (100) and (010) and around them the net of hk0 points.

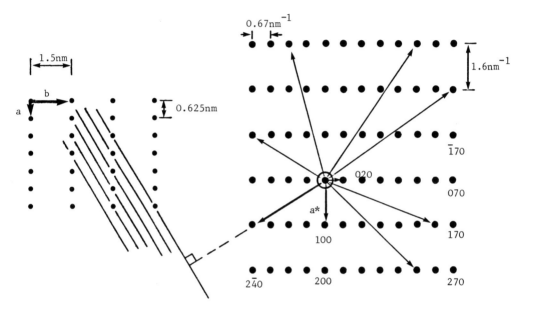

Again, to demonstrate the properties of the reciprocal lattice, the
1$\bar{4}$0 vector is drawn to show that it is perpendicular to the (1$\bar{4}$0) planes.

Index the rest of the net, and give in particular the indices of the
six vectors which are drawn.

The orthorhombic reciprocal lattice is now extended into three dimensions.

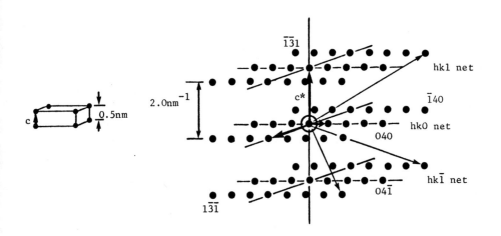

If we had begun this section with an orthorhombic lattice we might
have created the impression that a* is parallel to a and so on.
This happens to be the case in this orthorhombic lattice, but it is not
true in general.

Let us restate the relationship.

a* is perpendicular to (100)

b* is   "   (010)

c* is   "   (001)

Give the indices of the three reciprocal lattice vectors shown on the
diagram.

Complete the indexing of the reciprocal lattice nets
on the right.

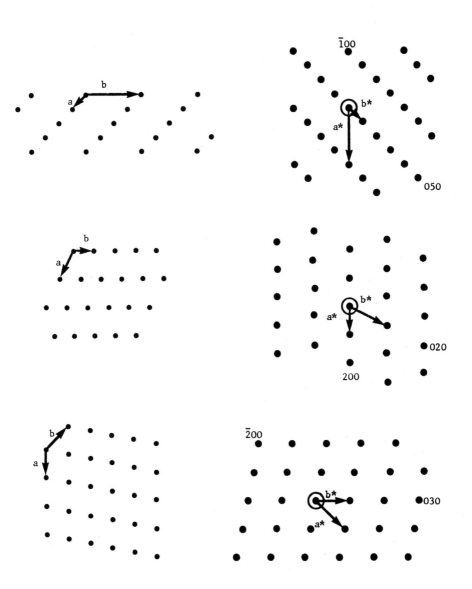

These reciprocal lattice nets pass through the origin of a cubic reciprocal lattice.

$\bar{2}00$

$\bar{1}\bar{1}\bar{1}$ $\bar{1}00$ $\bar{1}11$

$0\bar{1}\bar{1}$ $0\underset{a*}{\downarrow}$ $01\bar{1}$ $02\bar{2}$

100 111 122

200

Complete the indexing of each net.

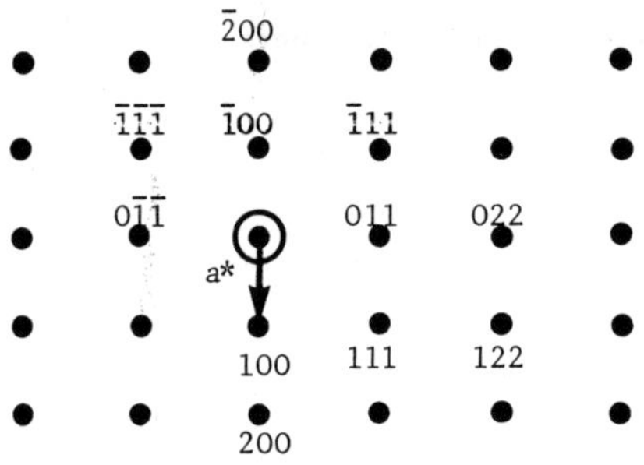

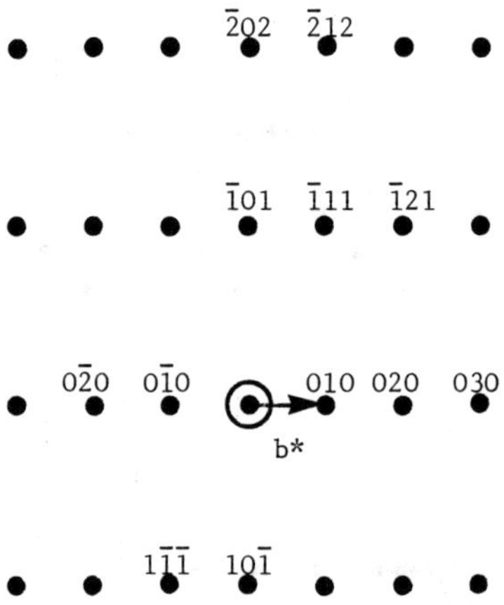

$\bar{2}02$ $\bar{2}12$

$\bar{1}01$ $\bar{1}11$ $\bar{1}21$

$0\bar{2}0$ $0\bar{1}0$ $\underset{b*}{\rightarrow}$ 010 020 030

$1\bar{1}\bar{1}$ $10\bar{1}$

Notice how the indices add, vector-fashion.

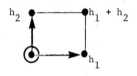

h_2 — $h_1 + h_2$

h_1

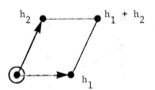

h_2 — $h_1 + h_2$

h_1

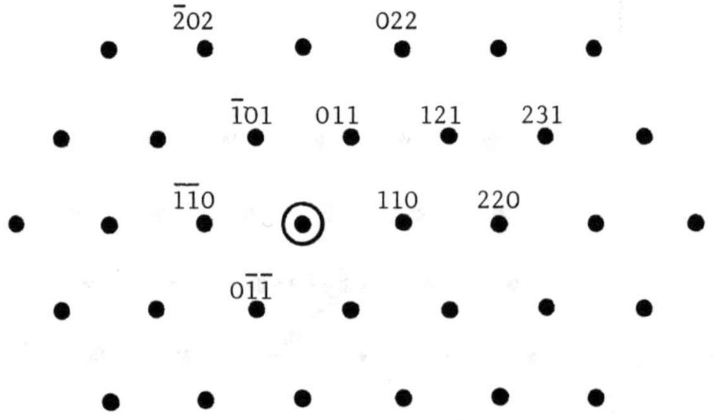

$\bar{2}02$ 022

$\bar{1}01$ 011 121 231

$\bar{1}10$ 110 220

$0\bar{1}\bar{1}$

We often use the terms "net" and "row" of reciprocal lattice points, and their meanings are obvious. On this hk0 net of a reciprocal lattice the broken lines show the rows hh0, h2h0, h3h0, 0k0 and h3h̄0.

Complete the indexing of the net, and identify the rows.

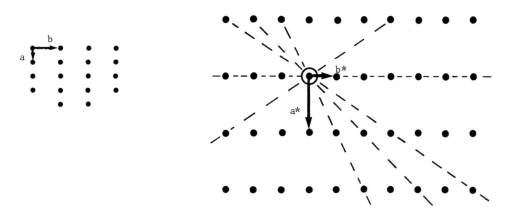

Do the same thing for the hexagonal reciprocal lattice net on the right below. The rows are hh2h̄0, 0kk̄0, h2h̄h0 and hh̄00, using the Miller-Bravais indices which were explained in Section 2.

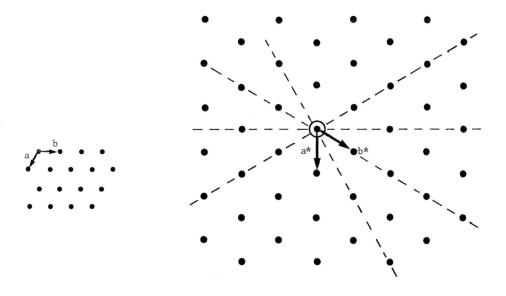

Before going further, we discuss briefly two aspects of the relation between the real and reciprocal lattices and then introduce our next move.

Vectors $\underline{a}$, $\underline{b}$, $\underline{c}$ and $\underline{a}^*$, $\underline{b}^*$, $\underline{c}^*$

We have constructed $\underline{a}^*$ perpendicular to the planes (100). These planes inevitably contain $\underline{b}$ and $\underline{c}$, and so we can say

	$\underline{a}^*$	is perpendicular to	$\underline{b}$ and $\underline{c}$
Also	$\underline{b}^*$	"	"	$\underline{c}$ and $\underline{a}$
and	$\underline{c}^*$	"	"	$\underline{a}$ and $\underline{b}$

The origin

Notice that we always begin the construction of the reciprocal lattice by setting up an origin 000 and then the axes. This origin is fixed and is central to the whole lattice, and it is the point to which the indices of all other points are related. The real lattice, on the other hand, has no such fixed origin. Any lattice point in the real lattice can be taken as an origin, and all points are identical.

The reciprocal lattice differs in this fundamental way from the real lattice in that it is not repetitive (apart from having equally spaced points). Its origin remains the origin, and the 001 and 002 points, for example, have their own identity and cannot be interchanged. But in the real lattice each point has nothing, such as indices, to distinguish it from its neighbours.

The reciprocal lattice and the Bragg law.

We shall show that the reciprocal lattice gives us a simple and elegant way of picturing the reflection of x-rays (and electrons in the next section) by crystal planes. The Bragg concept of planes reflecting the beam is itself simple, but a crystal contains many sets of planes set at bewildering angles to one another, and it strains the imagination attempting to decide how all these planes will affect a beam entering the crystal. The problem will be solved neatly with the reciprocal lattice and the Ewald sphere.

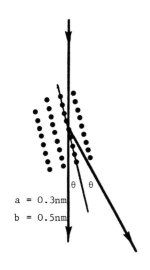

a = 0.3nm
b = 0.5nm

Bragg and the reciprocal lattice

The crystal on the left is reflecting
an x-ray beam from its (010) planes ;
d_{010} = 0.5nm, λ = 0.25nm, so θ = 14.5°.

Below, a sphere of radius $1/\lambda$ (i.e. $4nm^{-1}$)
has been drawn around the crystal, and
the reciprocal lattice has been added
<u>with its origin where the beam leaves</u>
<u>the sphere.</u>

The reciprocal lattice is
tilted because the real lattice
is tilted; they must turn
together.

Notice that the <u>010 point</u>
<u>is on the sphere and</u>
<u>the beam is passing</u>
<u>through it.</u>   Whenever
a point touches the
sphere the Bragg law
for the planes is
satisfied, and a
reflected beam appears.

(From the origin to the
010 point is $1/d_{010}$
which is also
$\frac{2}{\lambda}$ sin θ)

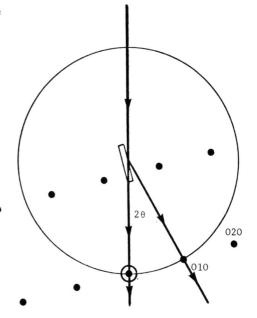

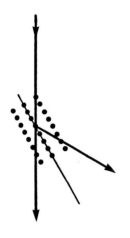

The same crystal on the left is now turned further, until $\theta = 30^o$, and the 020 planes are reflecting the beam. ($d_{020} = 0.25nm$)

Below, the reciprocal lattice has turned too , of course, <u>about its origin at the point where the beam leaves the sphere.</u> As we expect, the 020 point is on the sphere, and the reflected beam is passing through it.

Notice that the diagram also tells us something that we would not have suspected, that the $(\bar{1}30)$ and $(\bar{1}\bar{1}0)$ planes are also reflecting, since their points are on the sphere.

It soon becomes very easy to picture the reflected x-ray beams in this way, as being produced when reciprocal lattice points touch the sphere. Points touching a sphere are easier to imagine than planes satisfying the Bragg law.

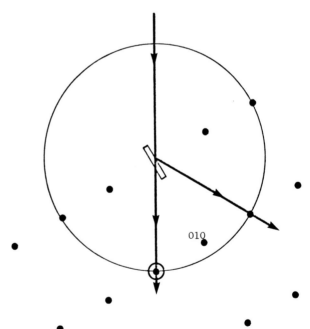

To show the usefulness of the Ewald sphere, as it is called, we take a
new crystal now, on the left, and ask, "Which planes, if any, will be
reflecting the beam?"

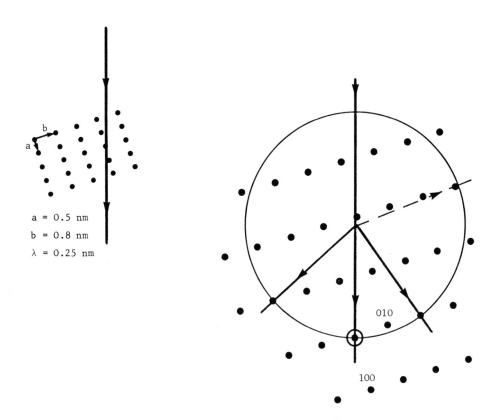

a = 0.5 nm
b = 0.8 nm
λ = 0.25 nm

To answer, draw the Ewald sphere of radius 4 nm^{-1} (i.e. 1/0.25) around the
crystal, and place the reciprocal lattice origin at the point where the
beam leaves the sphere.   (The reciprocal lattice is tilted to be lined up
with the real lattice.   One cannot tilt without the other.)

We see at a glance that the 020 and $\overline{12}0$ planes are reflecting and that the
$\overline{2}40$ planes are probably giving a faint beam, since they are nearly in the
correct position.

Now suppose the crystal were to be rotated slowly anticlockwise.
Which would be the next two beams to appear?   (To find out, rotate
the reciprocal lattice about its origin anticlockwise.)

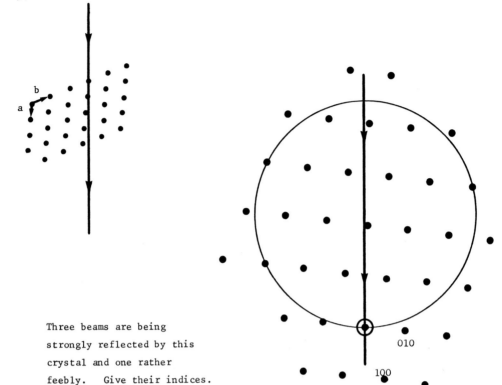

Three beams are being
strongly reflected by this
crystal and one rather
feebly.   Give their indices.

It would not be possible, by rotating this crystal, to obtain the $\bar{5}20$
and $\bar{5}30$ reflections (at the top of the diagram).   What change would we
have to make if we were determined the obtain these reflections?

On this and the previous few pages we have taken a 2-D view of the Ewald
sphere and ignored the third dimension.   From now on we shall bear in mind
that reciprocal lattice points above and below the page which touch the
sphere will also produce reflected beams which radiate, like all the others,
from the crystal at the centre of the sphere.

In this diagram, x-rays of short wavelength (MoKα, λ = 0.071 nm)
strike a crystal with a fairly large cubic unit cell (a = 1.25 nm)

The Ewald sphere is very large,
and only part of it
is shown, but clearly
the crystal reflects many
beams simultaneously.

All beams radiate
from here, where
the crystal is
placed.

Index these eight
reflected beams.
Take a* as pointing
down the page, b* to
the right and c* up out
of the page.

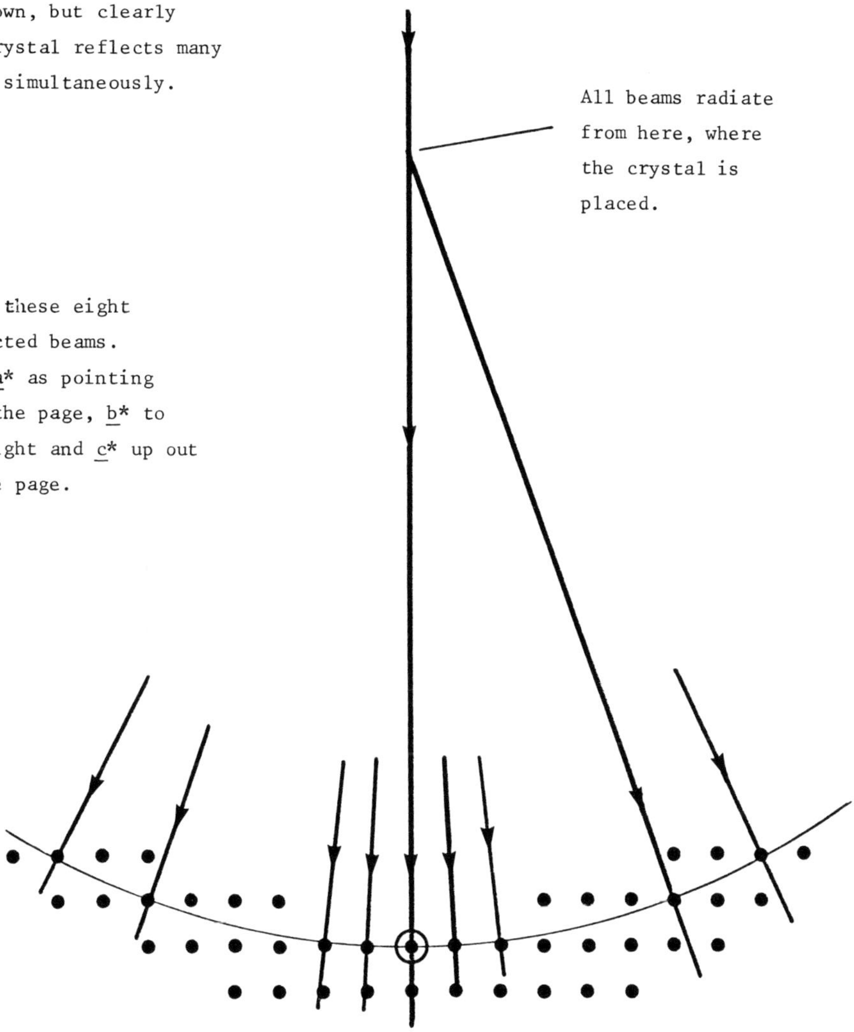

This diagram forms a useful link with Section 7, where electrons will be
reflected, and their wavelength is even shorter than that of these x-rays.

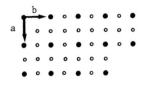

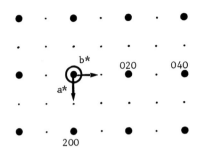

The cubic F or f.c.c. lattice
is in real space above, and
two nets of the reciprocal
lattice and a perspective
view of the lattice are on
the right.

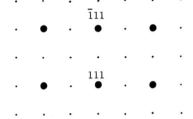

Complete the indexing of the
two nets, but read below first.

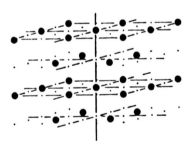

Many points are missing from this reciprocal lattice and marked with
faint dots, and if you have understood the previous five pages you may
realise why; they give no reflected beams.   The rest are shown in the
usual way, and these have h, k and ℓ odd or even.   Remember from
Section 5 that these are the only ones which give reflections.

This is a "weighted" reciprocal lattice, and we shall meet it again
in Section 7.

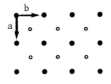

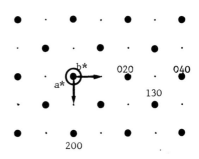

These reciprocal lattice nets
too are weighted, but this time
we have the cubic I or b.c.c.
lattice, and only points
with the sum h + k + ℓ even
are present.
The rest give no reflections
and are omitted.

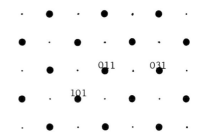

Complete the indexing of the
two nets.

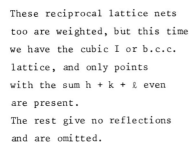

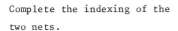

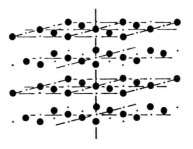

You will occasionally find it stated that the f.c.c. real lattice
has a b.c.c. reciprocal lattice and vice versa. This is not correct,
even though the diagrams on this and the previous page show what is
meant. The reciprocal cell is not centred, and the idea of centring
is simply not applicable to reciprocal space.

SECTION 7

Electron diffraction

Electrons, like x-rays, may be reflected (or diffracted) in
passing through a crystal, and they are governed by the Bragg
law.   In two respects they differ from x-rays.

The first is that they are charged particles and interact
strongly with the atoms of the crystal, so while the x-rays of
Section 5 passed through 0.5 mm of crystal, electrons would be
expected to penetrate less than a thousandth of that.

The second is that their wavelength is much shorter, and is
determined by their velocity.   The usual accelerating voltage
in an electron microscope (and that is where the diffraction
will take place) is 100kV, giving a wavelength about one fortieth
of the x-ray wavelength.

The relation between the reciprocal lattice and an electron
diffraction pattern is a particularly close one, so close in
fact that the pattern is identical to a net of the reciprocal
lattice.   With familiarity, one begins to look at the pattern
as a reciprocal lattice and almost forget the beams being reflected
by the real crystal.   An electron microscopist "sees" the
reciprocal lattice on the viewing-screen.

Diffraction and reflection of electrons

Here again we shall refer to "diffraction" patterns formed by beams
"reflected" by the crystal, an acceptable dual terminology which is
familiar to everyone.

But the section ends by touching upon a subject somewhat outside
the theme of this book, the production of an image by an electron
microscope, and perhaps at this point, appropriately, the term
reflection begins to outgrow its usefulness.   It serves our purpose
here, but a lens can produce images of things other than crystalline
materials, objects which have no recognisable planes within them
to give reflected beams, and a comprehensive account of image
formation requires an approach through a study of diffraction.

Imagine the electron beam
striking an extremely thin
crystal and the reflected beams
forming a pattern of spots on a
photographic film.

The true experimental set-up
is more complex than this, and
it will be explained at the
end of the section, but even
after the explanation you will
keep this picture in mind
whenever you see an electron
diffraction pattern.

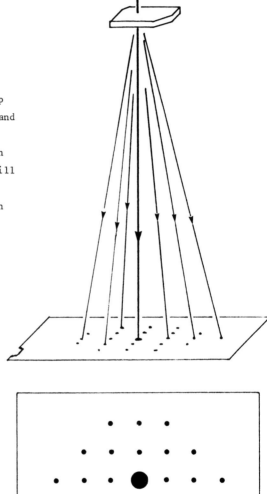

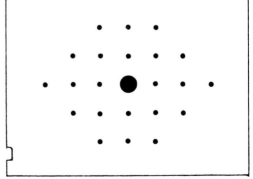

Think of the Ewald sphere around the crystal and the beams radiating through the reciprocal lattice points which touch it.

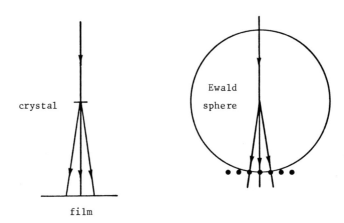

crystal

film

Ewald sphere

But our electrons will be accelerated through 100kV, giving them a wavelength of 0.0037nm, and $1/\lambda$ is 270nm^{-1}. So on the scale used in Section 5 the Ewald sphere radius is 270cm, about 9 feet. Part of a 9ft radius sphere is drawn below, and clearly it slices through many points in the centre of the reciprocal lattice net, so the pattern on the film will be a copy of the net. Notice also that the beams actually diverge only about 1° or so.

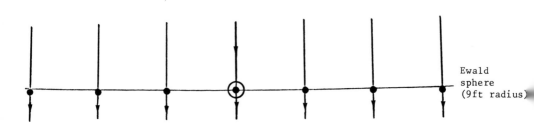

Ewald sphere (9ft radius)

Complete the indexing of these electron diffraction patterns
just as though they were weighted reciprocal lattice nets (i.e. no
reflection, no reciprocal lattice point).

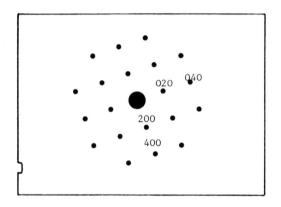

This hk0 pattern of
aluminium (fcc)
has indices all odd
or all even.

On this hk0 pattern
from tungsten (bcc),
h + k + ℓ is
always even.

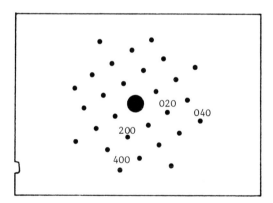

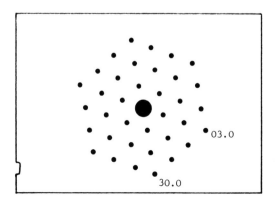

This basal hexagonal
pattern from zinc has all
the hk.0 reflections present.

Measuring the lattice parameter of aluminium

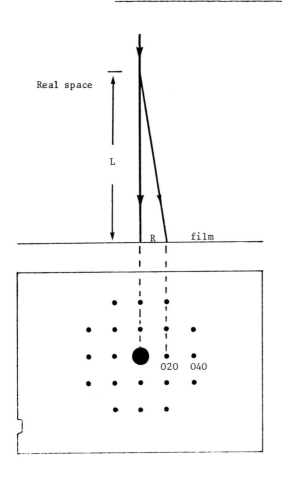

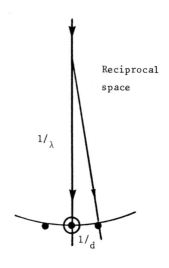

Comparing the two diagrams

$$\frac{R}{L} = \frac{1/d}{1/\lambda}$$

So $d = \frac{\lambda L}{R}$

For the 020 spot on this film, R = 9.5mm

Also, λ = 0.0037nm and L = 510mm

$$d_{020} = \frac{0.0037 \times 510}{9.5} = 0.199nm$$

$$a = d_{020} \sqrt{4} = 0.40nm$$

We shall not give λ and L separately from now on, but simply λL, which is known as the camera constant.

For each of these films, use $d = \lambda L/R$ to obtain d, and then calculate the lattice parameter. The camera constant λL is 1.90nm.mm for both films.

$R_{040} = 24$ mm

$(a = d_{040} \sqrt{16})$

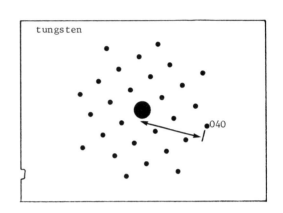

$R_{030} = 24.5$mm

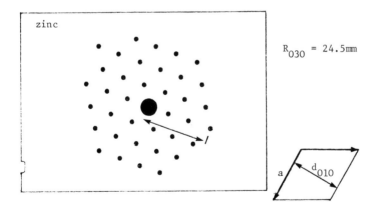

For a hexagonal unit cell,

$$a = d_{010} \times \frac{2}{\sqrt{3}}$$

$$(\text{and } d_{010} = 3 \times d_{030})$$

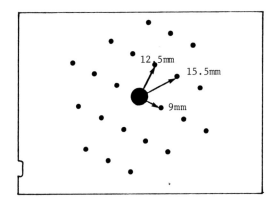

This pattern given by vanadium is to be indexed. We know that the metal is bcc and the lattice parameter is 0.302nm

From this, we calculate the d-values of the low-index planes.

hkl	d
110	0.21 nm
200	0.15 nm
211	0.12 nm
220	0.11 nm
310	0.10 nm

We know also that λL for this film is 1.90 nm.mm. From the values of R, calculate d for the three chosen spots on the film, using $d = \lambda L/R$.

Compare the three d-values with the above table to obtain the indices on the far left. Then make them consistent, vector-fashion, as on the right.

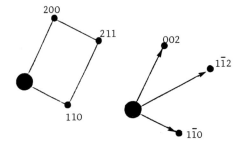

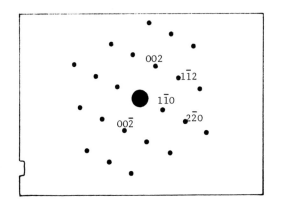

Complete the indexing of the pattern.

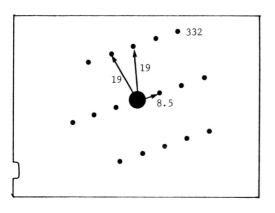

Index these two patterns
given by molybdenum
(bcc, and a = 0.315 nm)

As before, λL = 1.90 nm.mm

From the values of R (in mm) given on the films, calculate d from
d = λL/R, and index the spots by comparing d with the known values
of d for molybdenum given in the table.

hkℓ	d
110	0.22nm
200	0.16nm
211	0.13nm
220	0.11nm
310	0.10nm

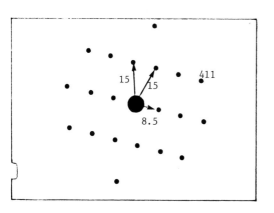

One spot on each film has been
indexed to provide you with a
check on your solution, but you
may, of course, in making your
indices consistent, arrive at
other, but equally valid, indices.
For example $\bar{1}4\bar{1}$ instead of 411.

These are diffraction patterns from copper, which has the cubic F
lattice (fcc) and a = 0.361nm.   Again, the camera constant
λL is 1.90 nm.mm.

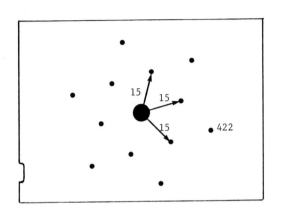

Index the patterns by, as before,
comparing the d-values obtained
from the R-values given on the
films (in mm) with those
calculated from the lattice
parameter and given below.
(Only indices all odd or even
are included, of course.)

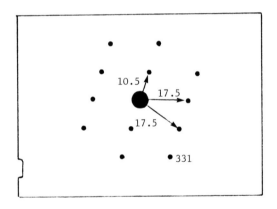

hkℓ	d
111	0.21nm
200	0.18nm
220	0.13nm
311	0.11nm
222	0.10nm

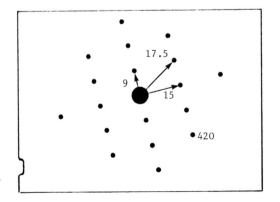

(The check indices such as
420 may appear as $\bar{2}04$, for
example, in your pattern).

Powder patterns and measurement of λL

If the electron beam, instead of striking a single crystal, strikes a polycrystalline specimen, one which is formed by deposition from the vapour for example, each crystallite gives its own pattern of spots. The crystallites being haphazardly arranged, a pattern of rings, probably spotty rings, is formed on the film.

Each crystallite has its own reciprocal lattice, and the points of the lattices will lie on spheres around a common origin. The concentric spheres cut the Ewald sphere in rings, which then give the rings on the film.

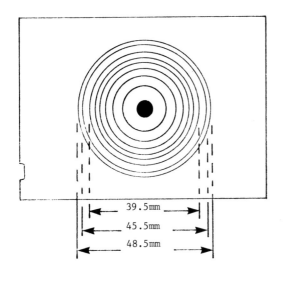

This pattern is of $T\ell C\ell$, cubic P with a = 0.383nm.

The indices of the rings are obviously, from the inside, 100, 110, 111, 200 and so on.

Index the rings. Calculate d for each of the three outer ones (from the given a–value). Obtain λL from d = λL/R. (R is the <u>radius</u>)

λL is invariably found in this way, never by finding λ and L separately.

108

Calculate λL for each of these polycrystalline patterns.   The first is
again thallium choride, with the cubic P lattice, and there are no
absent rings (except No.7 of course).   The next two, gold and aluminium,
are cubic F, with irregularly spaced rings.   (Remember the cubic F
patterns in Section 5.)

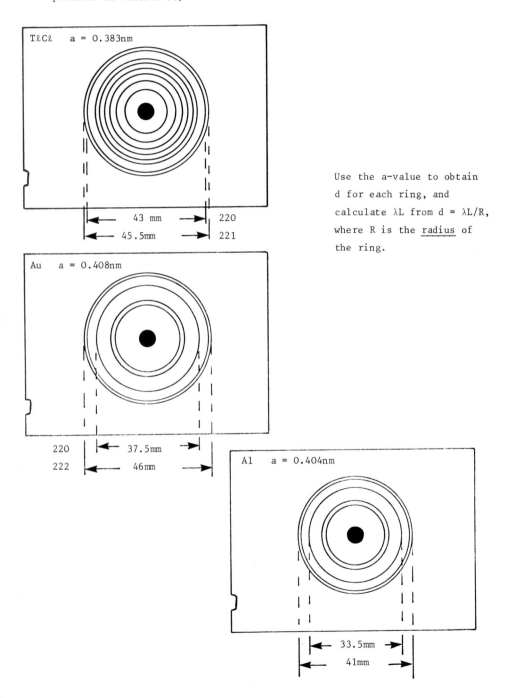

TℓCℓ   a = 0.383nm

|← 43 mm →| 220
|← 45.5mm →| 221

Use the a-value to obtain
d for each ring, and
calculate λL from d = λL/R,
where R is the radius of
the ring.

Au   a = 0.408nm

220 |← 37.5mm →|
222 |← 46mm →|

Al   a = 0.404nm

|← 33.5mm →|
|← 41mm →|

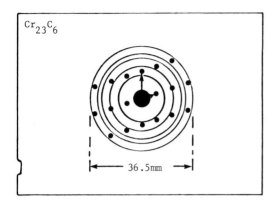

The spots to be indexed are given by the cubic carbide $Cr_{23}C_6$ which is fcc (a = 1.06nm). The R-values of the two spots arrowed are 5.75mm and 9.75mm

The rings are of $T\ell C\ell$ (a = 0.383nm) used as a standard to determine λL, and the diameter of the 210 ring is given.

It will be necessary, of course, to compile a table of d-values for $Cr_{23}C_6$. About twelve with the lowest indices should be sufficient.

The next pattern to be indexed is of the fcc carbide TiC (a = 0.432nm), and gold (a = 0.408nm) has been evaporated on to the specimen to provide a standard. The indices of the rings are 111, 200, 220, 311, 222. The R-values of the three arrowed spots are 10.5mm, 20mm and 17mm.

Again, a table of d-values, this time of TiC, is needed.

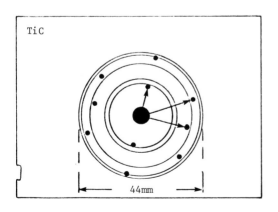

110

The electron microscope

We said at the outset that the
experimental arrangement we have
pictured so far, shown top right,
is an oversimplification (but a
useful one), and we must now
explain matters fully.

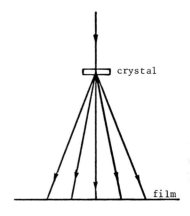

The specimen will in fact be in an
electron microscope and illuminated
by a very narrow beam of electrons
about 10μm diameter.   Our second
diagram, on a larger scale than the
first, shows it giving reflected
beams, just as we have already
described, and the beams just
leaving the crystal.

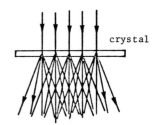

For clarity, five of the beams are drawn here separately.   The angles
of deflection (2θ) are necessarily exaggerated, since they are only
about 1°.   (Remember the 9ft Ewald sphere.)

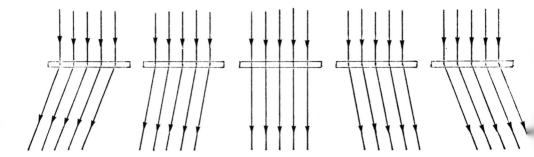

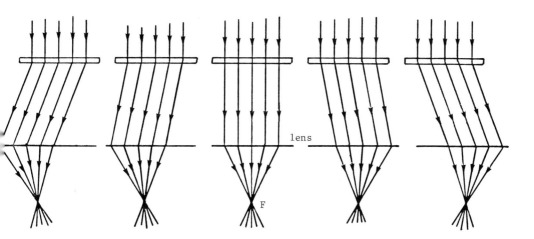

The beams now pass through the
first of the microscope's three
lenses, the objective, and each is
focused to a point in the focal plane,
and it is here that the complete
diffraction pattern is produced.

This second diagram shows just three
of the beams giving their spots in
the focal plane.

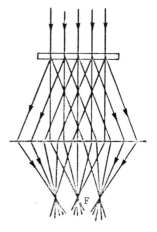

However, the focal length of the objective
is about 3mm, and since the beams are deflected
about 1°, the whole pattern is only about 0.1mm
across. The remaining two lenses are needed
to magnify the pattern, as we shall see.

This is the complete microscope, with
the objective giving the diffraction
pattern and the next two lenses each
giving a real, magnified image of it.
The final pattern is like the ones
we have used throughout, all of which
were life-size.

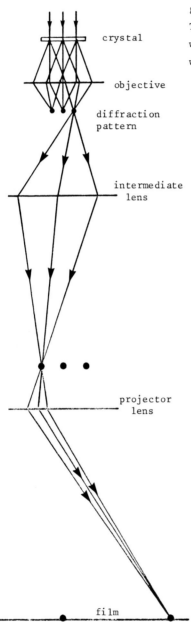

crystal

objective

diffraction
pattern

intermediate
lens

Notice that the very top of this
diagram is like the one with which
this section opened, and the
complications below it can be
forgotten in indexing a pattern.

projector
lens

Our explanation of the diffraction
pattern is now complete, but we
cannot abandon the microscope
at this stage.

film

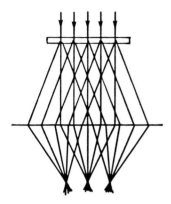

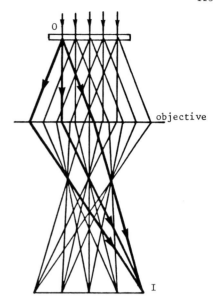

objective

I

Once the diffraction pattern has
been formed (above), the electrons
carry on and produce an image of
the crystal (top right).
Forming images is, after all,
the purpose of an electron
microscope.
The three thicker lines show
electrons spreading from an
object point O and converging
to an image point I.

The lower diagram shows the most
useful way of seeing how an image is produced.   Electrons spread out
from the diffraction spots to cover the area of the image, and it is the
addition of these beams which builds up the image.   One beam alone
would give uniform illumination, but together, giving constructive and
destructive interference, they produce the details of the image.

Understand that <u>the diffraction pattern is an inevitable stage in</u>
<u>the production of the image.</u>

So the objective gives, first, a diffraction pattern and, second, an image of the crystal. A magnified image of either of these can be provided by the other two lenses.

On the left below, the lenses give an image of the diffraction pattern, and on the right they give an image of the crystal. This is achieved by changing their focal lengths, and is made possible by the fact that the lenses are magnetic fields. Their strength can be varied simply by changing the current in the electromagnets.

The electron microscope is known generally for its very high resolving power, but this ability to show us both the crystal itself and its diffraction pattern is a further reason for its immense scientific value.

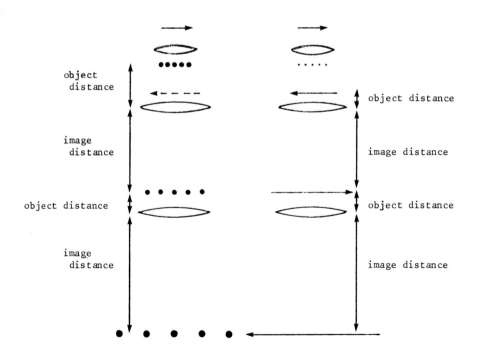

ANSWERS TO PROBLEMS

The problems and answers are not numbered, but the answers are arranged to correspond with the arrangement of problems on the page.

p6   rectangular   centred rectangular
      square   oblique
      rectangular   square

p7   square
      oblique   hexagonal   centred rectangular
      rectangular   square   oblique
      rectangular   hexagonal   centred rectangular

p9   square (4)   oblique (2)
      centred rectangular (2 atoms per point, 4 per unit cell)
      hexagonal (2)   rectangular (2)

p10   oblique (2)   oblique (2)
              rectangular (3)
      centred rectangular (3)   hexagonal (1)

p26 $(2\bar{1}0)$ or $(\bar{2}10)$ After this, Miller indices
      $(1\bar{1}0)$ or $(\bar{1}10)$ (230) or $(\bar{2}\bar{3}0)$ will be given in one
      $(4\bar{1}0)$ or $(\bar{4}10)$ $(1\bar{3}0)$ or $(\bar{1}30)$ form only.

p27 $(1\bar{3}0)$ (230) (130) p30 (122) $(1\bar{2}2)$ p31 $(1\bar{2}\bar{2})$ (122)
      (200) (040) $(3\bar{1}0)$ (320) $(3\bar{2}0)$ $(32\bar{1})$ (321)
      $(3\bar{2}0)$ $(1\bar{1}0)$ $(3\bar{4}\bar{2})$ (342)
      (150) $(1\bar{5}0)$ (140)
p32 (322) (122) $(31\bar{2})$ p33 (322) (314)
      (311) $(31\bar{1})$ $(32\bar{1})$ (040) (040)
      $(22\bar{2})$ (112) (221) $(6\bar{1}2)$ $(3\bar{1}8)$

p34 $(21\bar{3})$ (503) (221) p37 $(3\bar{2}\bar{1}0)$ $(10\bar{1}0)$
      $(50\bar{1})$ $(1\bar{1}3)$ $(1\bar{1}\bar{1})$ $(1\bar{2}10)$ $(31\bar{4}0)$
      (311) (216) $(2\bar{2}2)$ $(2\bar{3}10)$
      (014)

116

p38

(235)	(122)	Repeat these indices with
(253)	(212)	a bar over the h column,
(325)	(221)	then with a bar over the
(352)		k column and finally over
(523)		the ℓ column.
(532)		

---

(120)	($\bar{1}$20)	(220)	⟨200⟩
(102)	($\bar{1}$02)	(202)	(020)
(210)	($\bar{2}$10)	(022)	(002)
(201)	($\bar{2}$01)	($\bar{2}$20)	
(012)	(0$\bar{1}$2)	($\bar{2}$02)	
(021)	(0$\bar{2}$1)	(0$\bar{2}$2)	

p39 $[3\bar{1}0]$ $[\bar{1}\bar{2}0]$ $[\bar{2}\bar{1}0]$ $[\bar{1}20]$ $[120]$ $[320]$

p40   Upper section   Lower section

$[210]$	$[1\bar{2}0]$	$[\bar{6}10]$		$[310]$	$[\bar{1}20]$	$[\bar{6}\bar{1}0]$
$[\bar{2}10]$	$[\bar{3}\bar{1}0]$	$[5\bar{4}0]$		$[\bar{3}10]$	$[410]$	$[6\bar{1}0]$
		$[100]$				$[410]$
$[\bar{1}20]$	$[\bar{3}20]$	$[\bar{1}30]$		$[\bar{2}30]$	$[\bar{2}10]$	$[010]$

p41   Reading the arrows in their order around the diagram

$[\bar{2}\bar{1}\bar{1}]$		$[\bar{1}00]$	$[\bar{1}1\bar{1}]$
$[\bar{1}\bar{1}\bar{1}]$			$[01\bar{1}]$
$[\bar{1}\bar{1}0]$			$[\bar{1}11]$
$[0\bar{1}0]$			$[011]$
$[0\bar{1}1]$			$[111]$
$[1\bar{1}1]$	$[101]$	$[001]$	$[311]$

p42   Reading the arrows around the diagram

$[011]$ $[\bar{1}0\bar{1}]$ $[\bar{1}01]$
$[\bar{1}10]$ $[0\bar{1}1]$
$[01\bar{1}]$ $[\bar{1}\bar{1}0]$
                          $[0\bar{1}\bar{1}]$
                          $[101]$
$[110]$ $[10\bar{1}]$ $[1\bar{1}0]$

Double the number of directions given below by adding
the reverse of each one, e.g. $[\bar{1}\bar{1}\bar{1}]$ with $[111]$
                          and $[1\bar{1}\bar{1}]$ with $[\bar{1}11]$

$[111]$ $[112]$ $[\bar{1}12]$ $[1\bar{1}2]$ $[11\bar{2}]$
$[\bar{1}11]$ $[121]$ $[\bar{1}21]$ $[1\bar{2}1]$ $[12\bar{1}]$
$[1\bar{1}1]$ $[211]$ $[\bar{2}11]$ $[2\bar{1}1]$ $[21\bar{1}]$
$[11\bar{1}]$

$[123]$ Now place a bar over the h column
$[132]$ then a bar over the k column and then
$[213]$ over the ℓ column (making 24).
$[231]$ Finally, add the reverse directions
$[312]$ as stated above (making 48).
$[321]$

p43   The three unindexed directions are, from the left,
$[21\bar{3}0]$, $[14\bar{5}0]$ and $[\bar{1}5\bar{4}0]$. It is necessary to extend
the diagram in order to determine them.

p45 | Upper section, reading round the diagram

$a[\bar{3}20]$ $\frac{2a}{3}[\bar{3}\bar{1}0]$ $\frac{a}{2}[\bar{8}10]$ $a[\bar{2}30]$ Lower section

$\frac{a}{2}[\bar{1}30]$ $a[100]$ $\frac{3a}{2}[010]$ $\frac{a}{2}[\bar{1}10]$

$\frac{a}{3}[\bar{2}\bar{3}0]$ $\qquad$ $\frac{a}{2}[\bar{1}40]$ $\frac{a}{2}[\bar{4}\bar{1}0]$ $\frac{a}{2}[\bar{3}10]$ $\frac{3a}{2}[\bar{1}00]$ $\frac{a}{2}[\bar{1}\bar{3}0]$

$\frac{a}{2}[\bar{1}40]$ $\quad$ $a[100]$ $\frac{a}{2}[110]$ $a[140]$

p47

$\frac{a}{2}[\bar{1}30] = \frac{a}{2}[\bar{1}10] + a[010]$ $\qquad$ $2a[010] = \frac{a}{2}[\bar{1}20] + \frac{a}{2}[120]$

$2a[0\bar{1}0] = \frac{a}{2}[\bar{2}\bar{3}0] + \frac{a}{2}[2\bar{1}0]$ $\qquad$ $\frac{a}{2}[\bar{2}\bar{5}0] = 2a[0\bar{1}0] + \frac{a}{2}[\bar{2}\bar{1}0]$

$\frac{3a}{2}[010] = \frac{a}{2}[\bar{1}30] + \frac{a}{2}[100]$ $\qquad$ $\frac{a}{2}[1\bar{3}0] = \frac{3a}{2}[0\bar{1}0] + \frac{a}{2}[100]$

$a[100] = \frac{a}{2}[1\bar{2}0] + \frac{a}{2}[120]$ $\qquad$ $a[010] = a[1\bar{1}0] + a[\bar{1}20]$

$a[1\bar{1}0] = a[1\bar{2}0] + a[010]$ $\qquad$ $a[010] = \frac{a}{2}[\bar{4}10] + \frac{a}{2}[410]$

$a[2\bar{1}0] = 2a[100] + a[0\bar{1}0]$

p48 | No   p50 | Yes
| Yes | | Yes
| Yes | | Yes
| Yes | | Yes
| Yes | | No
| | | No
| | | Yes
| | | No

p51 | $[001]$ $[01\bar{1}]$ p52 | $(10\bar{1})$ $(\bar{1}30)$
| $[00\bar{1}]$ $[\bar{2}11]$ | | $(0\bar{1}1)$ (001)
| $[1\bar{1}0]$ $[\bar{1}01]$ | | $(21\bar{4})$ $(8\bar{6}5)$
| $[\bar{1}02]$ $[12\bar{2}]$ | | (221) $(1\bar{1}4)$

p53   Yes   Yes   p54 Yes   No   p55 90° 120°
      Yes   Yes   Yes   Yes 60° 30°
      No   Yes   No   No 90° 90°
      Yes   Yes   Yes   Yes 45° 135°

p59   cubic P   Cs at 000   Cl at $\frac{1}{2}\frac{1}{2}\frac{1}{2}$

      cubic P   Ca at $\frac{1}{2}\frac{1}{2}\frac{1}{2}$ Ti at 000   O at $\frac{1}{2}$00, 0$\frac{1}{2}$0, 00$\frac{1}{2}$

p60   cubic P   Al at 000   Ni at $\frac{1}{2}\frac{1}{2}$0, 0$\frac{1}{2}\frac{1}{2}$, $\frac{1}{2}$0$\frac{1}{2}$

      cubic P   Re at 000   O at $\frac{1}{2}$00, 0$\frac{1}{2}$0, 00$\frac{1}{2}$

            or   Re at $\frac{1}{2}\frac{1}{2}\frac{1}{2}$ O at $\frac{1}{2}\frac{1}{2}$0, 0$\frac{1}{2}\frac{1}{2}$, $\frac{1}{2}$0$\frac{1}{2}$

p61   W at 000   C at $\frac{2}{3}\ \frac{1}{3}\ \frac{1}{2}$

      W at $\frac{2}{3}\ \frac{1}{3}\ \frac{3}{4}$ and $\frac{1}{3}\ \frac{2}{3}\ \frac{1}{4}$ C at 000

      Ni at 000 and 00$\frac{1}{2}$ As at $\frac{2}{3}\ \frac{1}{3}\ \frac{3}{4}$ and $\frac{1}{3}\ \frac{2}{3}\ \frac{1}{4}$

p62   cubic F
      At each lattice point are Ca at $\frac{1}{4}\frac{1}{4}\frac{1}{4}$ F at 000 and 00$\frac{1}{2}$
                                 and Na at $\frac{1}{2}$00   Cl at 000

p63   cubic F
      C or S at 000 $\frac{1}{2}\frac{1}{2}$0   0$\frac{1}{2}\frac{1}{2}$ $\frac{1}{2}$0$\frac{1}{2}$
and C or Zn at $\frac{1}{4}\frac{1}{4}\frac{1}{4}$ $\frac{3}{4}\frac{3}{4}\frac{1}{4}$ $\frac{1}{4}\frac{3}{4}\frac{3}{4}$ $\frac{3}{4}\frac{1}{4}\frac{3}{4}$

p67 9.9° 14.9° 20.0° 30.9° 39.9° 50.4° 74.3°

p68   1.20nm 0.60nm 0.40nm 0.30nm 0.24nm 0.20nm

p74   Last six arcs   310, 311, 222, 320, 321, 400

p75   AlNi 0.289nm   AuZn 0.319nm

p77   BeCu 0.2698nm   BePd 0.2813nm   LiAg 0.3168nm

p78   LiF 0.400nm

p79   tungsten 0.315nm

p80   Cu 0.360nm   Ta 0.331nm   AgF 0.492nm

p84 $0\bar{1}2$ 121 $\bar{1}\bar{1}\bar{1}$

p85 $\bar{3}\bar{2}0$ $\bar{3}50$ $\bar{2}70$
        $\bar{1}\bar{4}0$ 160
        $1\bar{4}0$ 250

p86 $\bar{1}41$ $\bar{1}4\bar{1}$ $14\bar{1}$

p89   Upper section   Lower section

The 0k0 row lies across the page and the rows reading clockwise are

h3h0, h2h0, hh0, $h3\bar{h}0$

The h2hh0 lies across the page and the rows reading clockwise are

$0kk0$, $hh2\bar{h}0$, $h\bar{h}00$

p93   First $\bar{1}40$ then $\bar{3}40$, very soon followed by 030

p94   Strongly $\bar{1}\bar{2}0$, $\bar{3}\bar{1}0$, $\bar{3}40$
        Feebly $0\bar{1}0$
        To obtain $\bar{5}20$ and $\bar{5}30$, use shorter λ to increase the radius $(1/\lambda)$ of the Ewald sphere.

p95   From the left
        $\bar{2}80$, $\bar{1}\bar{6}0$, $0\bar{2}0$, $0\bar{1}0$, 010, 020, $\bar{1}60$, $\bar{2}80$

p103   tungsten 0.32nm   zinc 0.27nm

p107 λL = 3.1nm.mm

p108   TlCl pattern λL = 2.9nm.mm
        Au pattern λL = 2.7nm.mm
        Al pattern λL = 2.4nm.mm

p109   The indices of the two arrowed spots on the $Cr_{23}C_6$ pattern are 200 (5.75mm) and 311(9.75mm).
On the TiC pattern they are 111 (10.5mm), 311 (20mm) and 220 (17mm).